From The Reviews of THE PURPLE CLOUD . . .

"This masterpiece should live as long as the Odyssey."

—Jules Claretie, *Revue des Deux Mondes*

"M.P. Shiel is a writer of imperial imagination."

—Rebecca West

"He tells of a wilder wonderland than Poe ever dreamed of."

—Arthur Machen

"What a man! What an imagination!"

—Carl Van Vechten

"The greatest writer of sensational fiction of his time."

—The Irish Statesman

"Had Carlyle shared Coleridge's penchant for laudanum, he might have written thus."

—The English Review

"He is not to be touched, because there is no one else like him."

—Hugh Walpole

"A GENIUS DRUNK WITH THE HOTTEST JUICES OF OUR LANGUAGE."

—New York Post

THE PURPLE CLOUD

M. P. Shiel

A Warner Communications Company

Orig. pub. 1901

WARNER PAPERBACK LIBRARY EDITION

First Printing: August, 1963
Second Printing: April, 1966
Third Printing: November, 1973
Fourth Printing: January, 1974

This Warner Paperback Library edition is published by arrangement with The Vanguard Press, Inc.

Cover illustration by Chuck Sovek

Warner Paperback Library is a division of Warner Books, Inc., 75 Rockefeller Plaza, New York, N.Y. 10019.

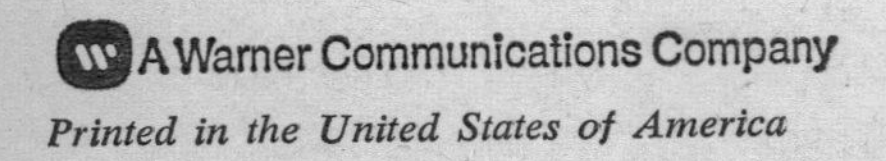

Printed in the United States of America

INTRODUCTION

In May of this year the writer received as noteworthy a packet of papers as it has been his lot to examine—from a friend, Dr. Arthur Lister Browne, M.A., F.R.C.P.—consisting of four note-books, crowded with those giddy shapes of "shorthand," whose *ensemble* resembles startled swarms hovering on the wing—scribbled in pencil, and without vowels: so that their deciphering has been no holiday. The letter also which accompanied them was pencilled in shorthand; and this, together with the note-book marked "III," I now publish.

The following is Browne's letter:

"Dear Old Chap,—I have just been lying thinking of you, wishing that you were here to give one a last squeeze of the hand before I—— '*go*': for going I am. Four days ago I felt a soreness in the throat, so, passing by old Johnson's surgery at Selbridge, I asked him to have a look at me, and when he muttered something about membranous laryngitis it made me smile, but by the time I reached home I was hoarse, and not smiling: before night I had dyspnœa and laryngeal stridor. So I wired to London for Morgan, and, between him and Johnson, they have been opening my trachea, and singeing my inside with chromic acid and the cautery; but I am too old a hand not to know what's what: the bronchi involved—*too far*. Morgan is still, I believe, fondly longing to add me to his successful-tracheotomy statistics, but prognosis was always my strong point, and the small consolation of my death will be the beating of a specialist up his own street. So we shall see.

"I have been arranging some of my affairs this morning, and remembered these note-books—intended letting you have them months ago, but you know my habit of putting things off, and, then, the lady was living from whom I took down the statements: now she is dead, and, as a writing man, and a man, you should be interested, if you can contrive to decipher.

"I am under morphia at present, propped up in a nice little state of languor, and, as I am able to write, will tell you something about her: her name Mary Wilson; thirty when I met her, forty-five when she died; fifteen years of her. Do you know much about the philosophy of the hypnotic trance? That was the relation between us—hypnotist and subject. She had been under another man before my time, suffered from *tic* of the fifth nerve, had had most of her teeth drawn before I saw her, and an attempt had been made to wrench out the nerve on the left side by external scission. But it had made no difference: the clock of hell tick-tacked in that poor woman's jaw, and it was a mercy that ever she dropped across *me*: my organization was found to possess easy control over hers, and with a few suggestions I could expel her Legion.

"Well, you never saw anyone so singular as my friend, Miss Wilson: medicine-man as I am, I could never behold her without a sort of shock: she so suggested what we call 'the *other* world,' some odor of the worm, ghost more than woman! And yet I can hardly convey to you the why of this, except by dry details as to the contours of her lofty forehead, meager lips, pointed chin, ashen cheeks. She was lank and deplorably emaciated, her whole skeleton, except the femurs, being visible, her eyes of the bluish hue of cigarette-smoke or quinine-solution made fluorescent by X-rays, and they had the strangest, feeble, unearthly gaze, which at thirty-five her wisp of hair was white.

"She was well-to-do, lived alone in old Wooding Manor-house, five miles from Ash Thomas; and I, 'beginning' in these parts at the time, soon took up my residence at the manor, she insisting that I should devote myself to her alone.

"Well, I found that, in the state of trance, Miss Wilson possessed remarkable powers: not peculiar to herself in *kind*, but so reliable, exact, far-reaching, in degree. Any tyro in psychical science will now sit and discourse about the reporting powers of the mind in the trance-state—a fact which Psychical Research only after endless investigation admits to be scientific, but known to every old crone in the Middle Ages; but I say that Miss Wilson's powers were '*remarkable*,' because I believe that, *in general*, the powers manifest themselves more particularly with regard to space, as distinct from time, the spirit roaming in the present, travelling over a plain; but Miss Wilson's gift was special in this, that she travelled all ways, and easily in all but one, east, west, up, down, in the past, the present, and the future.

"This I discovered gradually. She would emit a stream of sounds—I can hardly call it *speech*—murmurous, guttural, mixed with puffy breath-sounds of the languid lips, this accompanied by an intense contraction of the pupils, absence of the knee-jerk, rigor, a rapt and arrant expression; and I got into the habit of sitting long at her bedside, fascinated by her, trying to catch the import of that visionary language which came croaking from her throat, puffing and fluttering from her lips, until in the course of years my ear learned to discern the words; 'the veil was rent' for me, too; and I could follow somewhat the trips of her musing and wandering spirit.

"I heard her one day utter some words which were familiar to me: 'Such were the arts by which the Romans extended their conquests, and attained the palm of victory'—from

Gibbon's 'Decline and Fall,' which I could guess that she had never read.

"I said in a stern voice: 'Where are you?'

"She replied, 'Us are eight hundred miles above. A man is writing. Us are reading.'

"I may tell you two things: first, that in trance she never spoke of herself as 'I,' but, for some reason, in this *objective* way, as '*us*': 'us are,' she would say, 'us went,' though, of course, she was 'educated'; secondly, when wandering in the past she always represented herself as being '*above*' (the earth?), and higher the further back in time she went; in describing present events she felt herself '*on*,' while, as regards the future, she invariably declared that '*us*' were so many miles '*within*.'

"To her travels in this last direction, however, there seemed to exist fixed limits: I say seemed, meaning that, in spite of my efforts, she never, in fact, went far in this direction. Three, four thousand 'miles' were common figures on her lips in describing her distance 'above'; but her distance 'within' never got beyond sixty. Usually, she would say twenty, twenty-five, appearing in relation to the future to resemble a diver, who, the deeper he strives, finds a more resistant pressure, until at no great depth resistance becomes prohibition, and he can no deeper strive.

"I am afraid I can't go on, though I could tell you a lot about this lady. For fifteen years, off and on, I sat listening by her dim bedside, until at last my expert ear could detect the sense of her faintest exhalation. I heard the 'Decline and Fall' from beginning to end; and though some of her reports were the most frivolous stuff, over others I have hung in a horror of interest. Certainly, I have heard some amazing words proceed from those spirit-lips of Mary Wilson. Sometimes I could hitch her repeatedly to any scene or subject that I chose by the mere use of my will; at other times the flighty waywardness of her foot eluded me: she resisted—she disobeyed; otherwise I might have sent you, not four note-books, but twenty. About the fifth year it struck me that I should do well to jot down her more connected utterances, since I knew shorthand, and I did. . . . Note-book 'III' belongs to the eleventh year, its history being this: I heard her one afternoon murmuring in the intonation used when *reading*, asked her where she was, and she replied: 'Us are forty-five miles within: us read, another writes. . . .'

"But no more of Mary Wilson now: rather let us think a little of A. L. Browne—with a breathing-tube in his trachea, and Eternity under his pillow. . . ." (Dr. Browne's letter then continues on subjects of no interest here.)

(My transcription of the shorthand book "III" I now proceed to give, merely reminding the reader that the words form the substance of a document to be written, or to be motived (according to Miss Wilson), in that Future, which, no less than the Past, substantially exists in the Present—though, like the Past, we see it not. I need only add that the title, division into paragraphs, &c., have been arbitrarily contrived by myself for convenience.)

[*Here begins the note-book marked "III".*]

THE PURPLE CLOUD

Well, the memory seems to be getting rather impaired now. What, for instance, was the name of that parson who preached, just before the *Boreal* set out, about the wrongness of any more attempts to reach the North Pole? Forgotten! Yet four years ago it was as familiar to me as my own name.

Things which took place before the voyage seem to be getting a little cloudy in the memory now: I have sat here, in the loggia of this Cornish villa, to write down some sort of account of what has happened—God knows why, since no eye can ever read it—and at the very beginning I cannot remember the parson's name.

He was a strange sort of man surely, Scotchman from Ayrshire, big, gaunt, with tawny hair; used to go about London streets in shough and rough-spun clothes, a plaid flung from one shoulder, and once I saw him in Holborn with his rather wild stalk, frowning and muttering to himself. He had no sooner come to London and opened chapel (I think in Fetter Lane), than the little room began to be crowded; and when, some years afterwards, he moved to a big establishment in Kensington, all sorts of men, even from America and Australia, flocked to hear the thunder-storms that he talked, though certainly it was not an age prone to rage into enthusiasms over that species of pulpit prophet and prophecy. But this particular man undoubtedly did rouse the strong dark feelings that sleep in the heart: his eyes were pretty singular and powerful; his voice from a whisper ran gathering, like snowballs, and crashed, much like the pack-ice in commotion yonder in the North; while his gestures were as uncouth and gawky as some wild man's of the primitive ages.

Well, this man—what *was* his name?—Macintosh? Mackay? I think—yes, *that* was it! *Mackay,* Mackay saw fit to take offence at the fresh attempt to reach the Pole in the *Boreal;* and for three Sundays, when the preparations were nearing completion, fulminated against it at Kensington.

The excitement as to the Pole had at this time reached a pitch which can only be described as *fevered,* if this expresses the strange ecstasy and unrest which prevailed: for the scientific interest which men had felt in this unknown region was now, suddenly, a thousand times intensified by a new interest—a tremendous *money* interest.

And the new zeal had ceased to be healthy in its tone as the old zeal had been: for now the mean demon Mammon was having a hand in this matter.

Within the ten years preceding the *Boreal* expedition no less than twenty-seven expeditions had set out, and failed. . . .

The secret of which new rage lay in the last will of Mr. Charles P. Stickney of Chicago, that shah of faddists, supposed to be the richest individual who ever lived, who, ten years before the *Boreal* undertaking, dying, had bequeathed 175 million dollars to the man, of whatever nationality, who first reached the Pole.

Such the actual wording of the will—"*man who first reached*": and from this loose method of designating the person meant had immediately broken forth a prolonged heat of controversy in Europe and America as to whether or no the testator meant *the Chief* of the first expedition which reached, until it was finally decided on legal authority that the actual wording held good, that it was the individual, whatever his station in the expedition, whose foot first reached the 90th degree of latitude, who would have title to the "swag."

At all events, the furore had risen, I say, to the pitch of fever; and, as to the *Boreal* in particular, the progress of her preparations was minutely conned in the newspapers, everyone was an authority on her fitting, and she was in every mouth a bet, a hope, a joke, or a jeer: for now, at last, it was felt that success was near. So this Mackay had an interested audience, if a somewhat startled, and a somewhat cynical, one.

A lion-hearted man this must have been, after all, to dare proclaim a point-of-view so at variance with the mood of his time! One against four hundred millions, they bent one way, he the opposite, saying that they were wrong, all wrong! People used to call him "John the Baptist Redivivus": and without doubt he did suggest something of that sort. I suppose that at the time when he had the audacity to denounce the *Boreal* there was not a sovereign on any throne who, but for loss of standing, would not have been glad of a galley-post on board.

On the third Sunday night of his denunciation I was there in that Kensington chapel, and I heard him. And the wild talk he talked!—seemed like a man delirious with inspiration.

We all sat hushed, while the man's prophesying voice ranged up and down through all the modulations of thunder, from the hurrying mutter to the reverberant burst and hubbub: and those who came to scoff remained to wonder.

What he said was this: That there was some sort of Fate, or Doom, connected with the Pole in reference to the human race; that man's continued failure, in spite of continual effort, to attain proved this; and that this failure constituted a lesson—*and a warning*—which the race disregarded at its peril.

The North Pole, he said, was not so far away, and the difficulties in the way of reaching it were not, on the face of them, so great: human ingenuity had achieved a thousand things a thousand times more difficult; yet in spite of over half-a-dozen well-planned efforts in the nineteenth century, and of thirty-one in the twentieth, men had never really reached, though some had pretended to: always we had been balked, balked, by some seeming chance—some restraining Hand: and herein lay the lesson—*herein the warning*. Wonderfully like "the Tree of Knowledge" in "Eden," he said, was that Pole: the rest of the earth open and offered to man—but *That* persistently veiled and "forbidden"; as when a father lays a hand upon his son, with "Not here, my child; where you will—not here."

But persons, he said, were free to stop their ears, and turn a callous consciousness to the whispers and hints of Heaven; and he believed, he said, that the time was now near when we would find it absolutely in our power to stand on that 90th of latitude, and plant an impious foot on the head of this planet—as it had been given into the power of "Adam" to stretch an impious hand to the "Tree of Knowledge"; but, said he—his voice vaulting now to a prolonged proclamation of awful augury—as the abuse of that power had been followed in the one case by downfall prompt and cosmic, so, in the other, he warned the whole human crew to look out thenceforth for nothing from God but a grumbling heaven, and thundery weather.

The man's frantic sincerity, authoritative voice, savage gestures, could not but have their effect upon all—as for me, I declare, I sat as though a messenger from Heaven addressed me; but I believe that I had not yet reached home when the whole impression of the discourse had passed from me like water from a duck's back. No, the Prophet in the twentieth century was not a success: John Baptist himself, camel-skin and all, would have met with only tolerant shrugs. I dismissed Mackay from my mind with the thought: "Behind his age, I suppose."

But haven't I thought differently of Mackay since, my God . . . ?

* * *

Three weeks—about that—before that Sunday-night discourse, I was visited by Clark, the chief of the expedition—visit of friendship, I having then been established a year at 24, Harley Street, and, though under twenty-seven, had, I suppose, as *élite* a practice as any doctor in Europe.

Élite—but small: I was able to maintain my state, and move among the great; but now and again I would feel a pinch: just about then, in fact, I was only saved from embarrassment by the success of my book, *Applications of Science to the Arts*.

In the course of conversation that afternoon Clark said to me in his haphazard way: "Do you know what I dreamed about you last night, Adam Jeffson?—that you were with us on the expedition."

I think he must have seen my start: on the same night I

had dreamed the same thing; but not a word said I about it now. There was a stammer in my tongue when I answered: "Who? I? on the expedition?—wouldn't go, if I were asked."

"Oh, you would"—from Clark.

"I wouldn't. You forget that I am about to be married."

"Well, we need not discuss it, as Peters is not going to die. Still, if anything did happen to him, it is you I should come straight to, Adam Jeffson."

"Clark, you jest," I said: "I know little of astronomy or meteorological phenomena. Besides, I am about to be married. . . ."

"But what about your botany, my friend? *There's* what we should be wanting of you; and as for nautical astronomy, poh, a man of your scientific habit would pick all that up in no time."

"You discuss the matter gravely, Clark," I said, smiling: "such a thought would never enter—There is, first of all, my *fiancée*——"

"Ah, the all-important Countess, eh?—Well, but she, as far as I know the lady, would be the first to force you to go. The chance of stamping one's foot on the Pole does not occur to a man every day, my son."

"Talk of something else!" I said: "there is Peters. . . ."

"Well, of course, there is Peters. But, believe me, the dream I had——"

"Oh, your dreams!" I laughed.

Yes, I remember: pretended to laugh! but my secret heart knew, even *then*, that one of those crises was occurring in my life which, from my childhood, have made it the most extraordinary that any creature of the earth ever lived; and I knew that this was so, firstly because of the two dreams, and secondly because, when, Clark gone, I was drawing on my gloves to go to see my *fiancée*, I heard distinctly the old two voices; and one said: "Go not to see her now!" and the other: "Yes, go, go!"

The two voices of my life! One, reading this, would think that I mean merely two contradictory impulses—or else that I rave: for what modern man could comprehend how real-seeming were those voices, how loud, and how anon I could hear them contend within me with a nearness "nearer than breathing," "closer than hands and feet."

About the age of seven it happened first to me: I playing one summer evening in a pine-wood of my father's; half a mile away a quarry-cliff; and it seemed as if someone said inside of me "Take a walk toward the cliff," and as if someone else said "Don't go that way at all!"—whispers then, which gradually, as I grew up, swelled to cries of wrathful contention. I did go toward the cliff: and fell. Some weeks later, on recovering speech, I told my astonished mother that someone "had pushed me" over the edge, and that someone else "had caught me" at the bottom!

One night, somewhat before my thirteenth birthday, lying on a sofa, the notion visited me that my life must be of mighty importance to some-thing or things that I could not see; that two Powers, which hated each other, must be continually after me, one wishing to kill me, the other to keep me living, one wishing me to do so and so, the other to do the opposite; that I was not a boy like other boys, but a being separate, special, marked for—something. Already then I had notions, touches of mood, fugitive instincts, as occult and primitive, I verily believe, as those of the first man that stepped: so that such expressions as "Lord spake to So-and-so, saying" have never suggested any question in my mind as to how the voice was *heard*: I did not find it difficult to comprehend that originally men had more ears than two, as beasts and "mediums" have. nor should have been surprised to know that I, in these latter days, more or less resemble those primeval ones.

But not a creature, except maybe my mother, has ever dreamed me what I here state that I was: I seemed the ordinary boy of my day, bow in my "Varsity eight," cramming for exams., dawdling in clubs. When I had to select a profession, who could have suspected the battle that transacted itself in my breast, while my brain was careless—that conflict wherein the brawling voices brawled, the one: "Be a doctor," the other: "Be a lawyer, an artist—be *anything* but a doctor!"

A doctor I became; went to what had grown into the greatest of medical schools—Cambridge; and there it was that I came across a man named Scotland, who had an odd view of the world—was always talking about certain "Black" and "White" Powers, till it became absurd, and the men used to call him "Black-and-white-mystery-man," because one day when someone said something about "the black mystery of the universe," Scotland corrected him with "the black-and-white mystery."

Well I remember Scotland now—had rooms in the New Court at Trinity, and a set of us were generally there—the gentlest soul, with a passion for cats, and Sappho, and the Anthology, very short in stature, with a Roman nose, continually making efforts to keep his neck straight, and draw his paunch in. He used to vow that the universe was being furiously contended for by two Powers: that the White was the stronger, but did not find the conditions on our particular planet very favorable to his success, had got the best of it up to the Middle Ages in Europe, but since then had been slowly, stubbornly, giving way before the Black; and finally the Black would win—not everywhere perhaps, but *here*—would carry off, if no other planet, at least *this* one, for his prize.

Such was Scotland's doctrine, which he never wearied of

repeating; and while others heard him with mere toleration, little could they divine with what a burning of inward interest, I, cynically smiling there, imbibed his words. Most profound, most profound was the impression they made upon me.

* * *

But I was saying that when Clark left me, I was drawing on my gloves to go to see my *fiancée,* the Countess Clodagh, when I heard the two voices most clearly; and since sometimes the urgency of one or the other impulse is so overpowering, that there is no resisting it, so it was now with the one that bid me go.

I had to walk the distance between Harley Street and Hanover Square, and all that time it was as though something called at my ear: "Breathe no word of Clark's visit!" and another call: "Tell, hide nothing!"

It seemed to last a month; yet it was only some minutes before I was in Hanover Square, and Clodagh in my arms.

She was, in my opinion, the most superb of creatures, Clodagh—that haughty throat which seemed to be always scorning something just behind her left shoulder. Superb! but, ah—I know it now—a godless woman, Clodagh, a bitter heart.

Clodagh once confessed to me that her favorite character in history was Lucrezia Borgia, and when she saw my horror, immediately added: "Well, no, I am only joking!" Such was her duplicity: for I see now that she lived in the effort to keep hidden her heinous heart from me. Yet, now I think of it, how completely did Clodagh enthral me!

Our proposed marriage was opposed by both my family and hers: by mine, because her father and grandfather had died in lunatic asylums; and by hers, because forsooth, I was neither a rich nor a noble match. A sister of hers, much older than herself, had married a common country-doctor, Peters of Taunton, and this so-called *mésalliance* made the so-called *mésalliance* with me doubly detestable to her relatives. But Clodagh's passion for me was to be stemmed neither by their threats nor prayers. What a flame, after all, was Clodagh! Sometimes she scared me.

She was at this date no longer young, being by five years my senior, as also by five years the senior of her nephew, born from the marriage of her sister with Peters of Taunton, this nephew being Peter Peters, who was to accompany the *Boreal* expedition as doctor, botanist, and meteorological assistant.

On that day of Clark's visit to me I had not been seated five minutes with Clodagh, when I said: "Dr. Clark—ha! ha!—has been talking to me about the expedition—says that if anything happened to Peters, I should be the first man he would run to—has had an absurd dream. . . ."

The consciousness that filled me as I uttered these words was the *wickedness* of me—the crooked wickedness. But I could no more help it than fly.

Clodagh, standing at a window, holding a rose at her face. for quite a minute made no reply; I saw her sharp-cut, florid face in profile, steadily bent and smelling, till she said in her cold, rapid way: "The man who first plants his foot on the Pole will certainly be ennobled. I say nothing of the many millions. . . . I only wish I was a man!"

"I don't know that I have any special ambition that way," I rejoined: "I am happy in my warm Eden with my Clodagh."

"Don't let me think little of you!" she answered pettishly.

"Why should you, Clodagh? I am not bound to desire to go to the North Pole!"

"But you *would*, I suppose, if you could?"

"I might—I—doubt it. There is our marriage. . . ."

"Marriage indeed! It is the one thing to transform our marriage from a sneaking difficulty to a ten times triumphant event."

"If *I* personally were the first to stand at the Pole; but there are many in an——"

"For *me* you will, Adam——"

" '*Will*,' Clodagh?" I cried: "you say '*will*?' there is not the shadow of a chance——!"

"But why? There are still three weeks before the start. They say . . ."

She stopped.

"They say what?"

Now her voice dropped: "That Peter takes atropine."

Ah, I started then, she now moving from the window to sit in a rocking-chair, to turn the leaves of a book, without reading; and we were silent, she and I, I standing, looking at her, she drawing her thumb across the leaf-edges, and beginning again, contemplatively, until she laughed dryly a little—a dry, mad laugh.

"Why did you start when I said that?" she asked, reading now at random.

"*I!* I did not start, Clodagh! What made you think that I started? I did not start! Who told you, Clodagh, that Peters takes atropine?"

"He is my nephew: I should know. But don't look dumbfoundered in that absurd fashion: I have no intention of poisoning him in order to see you a multimillionaire, and a Peer of the Realm. . . ."

"My dearest Clodagh!"

"I easily might, however. He will be here presently—bringing Mr. Wilson for the evening." (Wilson was going as electrician of the expedition.)

"Clodagh," I said, "believe me, you jest in a manner which I don't find pretty."

"Do I really?" she answered with that haughty half-turn of

Peters poisoned ... by whom?

her throat: "then, I must be more exquisite. But, then, it is only a jest. Women are no longer admired for doing such things."

"Ha! ha! ha!—no—no longer admired, Clodagh! Oh, well, let us change this talk. . . ."

But now she could talk of nothing else—got from me that afternoon the history of the Polar expeditions of late years, how far each had reached, by what aids, why they had failed: and her eyes shone; she listened eagerly. Before this, indeed, she had been interested in the *Boreal,* knew the details of her outfitting, was acquainted with several members of the expedition; but now, suddenly, her interest seemed inflamed, my mention of Clark's visit having apparently set her scarlet with Arctic-fever.

The heat of her kiss as I freed myself from her embrace that day I still remember. I went home with a rather heavy heart.

Well, from the house of Dr. Peter Peters, three doors from mine on the opposite side of the street, his footman ran to knock me up that midnight with the news that Peters was ill; and when I hurried to his bedside I knew by the first glance at his gay deliriums and staring pupils that he was poisoned with atropine.

Wilson, the electrician, who had passed the evening with him at Clodagh's in Hanover Square, and was there, said to me: "What on earth is the matter?"

"Poisoned," I answered.

"Good God! Atropine, is it?"

"Don't be frightened: I think he will recover."

"Pretty certain?"

"Yes—that is, if he leaves off taking the drug, Wilson."

"What! it is he who has poisoned himself?"

I hesitated; but then said: "He takes atropine."

Three hours I remained there, and, God knows, toiled hard for his life: when I left him in the dark of the morning my mind was at rest.

I slept till 11 a.m., then hurried over again to Peters, in whose room were one of my two nurses, and Clodagh; and at once my beloved put finger to lip, whispering: "Sh-h-h! he is asleep. . . ." then came close to my ear, saying: "I heard the news early—am come to stay with him, till—the last. . . ."

We looked at each other some time—eye to eye, steadily, she and I; but mine dropped before Clodagh's. A word was on my mouth to say, but I said nothing.

Well, the recovery of Peters was not so steady as I had expected. At the end of the first week he was still prostrate; and it was then that I said to Clodagh: "Clodagh, your presence at the bedside somehow frets me—so unnecessary."

"Unnecessary certainly," she replied: "but I always had a genius for nursing, and a passion for watching the battles of the body. Why do you object?"

"Oh, I don't know. . . . This is a case that I dislike: I have half a mind to throw it to the devil."

"Then, do so."

"And you, too—go home, go home, Clodagh!"

"But *why?*—if one does no harm. In these days of 'the corruption of the upper classes,' and Roman decadence of everything, shouldn't every innocent whim be encouraged by you upright ones who strive against the tide? I find a sensuous pleasure in dabbling in drugs—like Helen, for that matter, and Medea, and Calypso, and the great antique women, who were all chemists. To study the human ship in a gale, and the slow drama of its foundering—And I want you to acquire the habit of letting me have my little way——"

Now she touched my hair with a lofty playfulness which soothed me; but even then I looked upon the rumpled bed, and saw that the man there was really very sick.

I have still a nausea to write about it! Lucrezia Borgia in her own age may have been heroic: but Lucrezia in this late century! One could retch up the heart. . . .

The man grew sick on that bed, I say. The second week passed, and when only ten days remained before the start of the expedition Wilson, the electrician, was one evening seated by Peters' bedside when I entered, at a moment when Clodagh was about to administer a dose to Peters; but, seeing me, she put down the medicine-glass on the night-table, and came toward me; and, as she came, I saw a sight which stabbed me: for Wilson took up the medicine-glass deposited by her, elevated it, looked at it, smelled into it; and he did it with a kind of light-fingered stealth; and he did it with an underlook, and a meaningness of expression, which, it seemed to me, meant mistrust. . . .

Meantime, Clark came each day. He had himself a medical degree, and about this time I called him in professionally, together with Alleyne of Cavendish Square, to consultation over Peters, who now lay in a semi-coma broken by passionate vomitings; and his condition puzzled us all. I formally stated that he took atropine—had originally been poisoned by atropine: but we saw that his present symptoms were scarcely atropine symptoms, but, it almost seemed, of some other vegetable poison or poisons, which we could not definitely name.

"Mysterious thing," Clark said to me when we were alone.

"*I* don't understand it," I said.

"Who are the two nurses?"

"Oh, highly recommended people of my own."

"At any rate, my dream about you comes true, Jeffson. It is clear that Peters is out of the running now."

I shrugged.

"I now formally invite you to join the expedition," says Clark: "do you consent?"

I shrugged again.

"Well, if that means consent," he said, "let me remind you that you have only eight days, and all the world to do in them."

This conversation occurred in the dining-room of Peters' house; and, as we passed through the door, I saw Clodagh gliding down the passage outside—rapidly—away from us.

Not a word I said to her that day about Clark's invitation; yet I asked myself repeatedly: Did she not know of it? Had she not *listened,* and heard?

However that was, near midnight, to my surprise, Peters opened his eyes, smiled, and by noon the next day his fine vitality, which so fitted him for an Arctic expedition, had rallied, he then leaning on an elbow, speaking with Wilson; and, except his pallor and strong stomach-pains, scarcely a trace left of his late closeness to death. For the pains I prescribed some quarter-grain tabloids of sulphate of morphia, and went away.

Now, David Wilson and I never greatly loved each other, and that very day he brought about a painful situation as between Peters and me, by telling Peters that I had taken his place in the expedition.

On which Peters, a touchy fellow, at once dictated a letter of protest to Clark: and Clark sent Peters' letter to me, marked with a great note of interrogation in red pencil.

Now, Peters' preparations were completely made, mine not, and he had five days in which to recover himself: I therefore wrote to Clark, saying that the changed circumstances of course annulled my acceptance of his proposal, though I had already incurred the inconvenience of negotiating with a *locum tenens.*

So this decided it: Peters was to go, I stay. The fifth day before the departure dawned, a Friday, the 15th of June, Peters now in an arm-chair, cheerful, but with a fevered pulse, and still the stomach-pains, I now giving him three quarter-grains of morphia a day. That Friday night, at 11 p.m., I visited him, and found Clodagh there, talking to him, he smoking a cigar.

Clodagh said: "I was waiting for you, Adam; didn't know whether I was to inject anything tonight. Is it Yes or No?"

"What do you think, Peters?" I said.

"Well, perhaps you had better give us another quarter," he answered: "there's still some trouble in the tummy off and on."

"A quarter-grain, then, Clodagh," I said.

As she opened the syringe-box, she remarked with a pout: "Our patient has been naughty! He has taken some more atropine."

I got angry at once. "Peters," I cried, "you know you have no right to be doing things like that without consulting me! Do that once more, and I swear I have nothing further to do with you."

Peters recovers then dies of poison; Clodagh's involvement.

"Rubbish," says Peters: "why all the unnecessary heat? A mere flea-bite—I felt that I needed it."

"He injected it with his own hand," Clodagh mentioned. . . . She was now standing at the mantelpiece, having lifted the syringe-box from the night-table, taken from its velvet lining both the syringe and the vial containing the morphia tabloids, and gone to the mantelpiece to melt one of the tabloids in a little of the distilled water there, her back turned upon us: and she was a long time. I was standing; Peters in his arm-chair, smoking; Clodagh talking about a Charity Bazaar which she had visited that afternoon.

She was long, yes, and the crazy thought arose in some corner of my soul: "Why is she so *long*?"

"Ah, that was a pain!" Peters said: "never mind the bazaar, Aunt—think of the morphia."

Suddenly an irresistible impulse seized me—to rush upon her, to dash syringe, tabloids, glass, and all, from her hands. I *must* have obeyed it—I was on the tip-top point of obeying—my body already leant; but in that moment a voice at the open door behind me said: "Well, how is everything?"

Wilson, the electrician, stood there: and with lightning swiftness I remembered an underlook of mistrust which I had once seen in his eyes. . . . Oh, well, I would not, could not! she was my love—I stood like stone. . . .

Clodagh went to Wilson with frank right hand, in her left being the fragile glass containing the injection; and my eyes, fastened on her face, saw it full of reassurance, of free innocence; so that I said to myself: "I must be mad!"

An ordinary chat began, while Clodagh turned up Peters' sleeve, and, kneeling, injected his forearm; when she rose, laughing at some remark of Wilson's, the drug-glass dropped from her hand, and her heel, by an apparent accident, trod on it. As she put the syringe among a number of others on the mantelpiece, she mentioned once more with that same pout: "The patient has been naughty, Mr. Wilson—has been taking more atropine."

"Not really?" said Wilson.

"Let me alone, the whole of you," answered Peters: "I ain't a child."

Those were the last intelligible words he spoke: died shortly before 1 a.m., poisoned by atropine, in spite of the morphia, the antidote of atropine, that he had in him.

From that moment to the moment when the *Boreal* bore me down the Thames all was a tumbled dream to me, of which hardly any detail remains in my memory: I remember how at the inquest I was called upon to prove that Peters had himself injected himself with atropine; and, this having been corroborated by Wilson, and by Clodagh, the verdict was in accordance.

And in all that chaotic hurry of preparation, two other things only, but those with distinctness now, I remember.

The first—and chief—is that turmoil of words which I heard at Kensington from that big-mouthed Mackay on the Sunday evening. What was it that enticed me, busy as I was, to that building that night? Well, perhaps I know.

There I sat and heard him: and most strangely have those words of his peroration impressed themselves upon my brain, when, dashing to a passion of prophecy, he proclaimed: "And as in the one case the abuse of that power was followed by downfall prompt and cosmic, so, in the other, I warn the whole human crew to look out henceforth for nothing from God but a grumbling heaven, and thundery weather."

And this second thing I remember in all that turmoil of doubts and flurries: that, as the *Boreal* moved down with the afternoon tide, a wire was put into my hand, a last word from Clodagh, who said only this: "Be first—for Me"; and I said then in myself: "The woman gave me of the tree, and I did eat."

* * *

The *Boreal* left St. Katherine's Docks in beautiful weather on the afternoon of the 19th of June, full of good hope, bound for the Pole.

All about the docks was one region of heads stretched out in innumerable vagueness, and down the river to Woolwich a continuous roaring and murmuring of bees droned from both shores to cheer our journey.

The expedition was partly a national affair, subvented by Government: and if ever ship was well-found it was the *Boreal*, which had a frame tougher far than any battleship's, capable of ramming some ten yards of drift-ice, and was stuffed with sufficient pemmican, cod-roe, fish-meal, and so on, to last us not less than six years.

We were seventeen, all told, the five Heads (so to speak) of the undertaking being Clark (our Chief), John Mew (commander), Aubrey Maitland (meteorologist), Wilson (electrician), and myself (doctor, botanist, and assistant meteorologist).

The idea was to get as far east as the 100°, or the 120°, of longitude; to catch there the northern current; to push and drift our way northward; and, when the ship could no further penetrate, to leave her (either three, or else four, of us, on ski), and with sledges drawn by dogs and reindeer make a dash for the Pole.

This had also been the plan of the last expedition—that of the *Nix*—and of others, the *Boreal* only differing from the *Nix* in being a thing of nicer design, of more exquisite forethought.

Our voyage was without incident up to the end of July, when we encountered a drift of ice-floes. On the 1st of August we were at Kabarova, where we met our coal-ship, and took in some coal for an emergency, liquid air being our proper motor; also forty-three dogs, four reindeer, and a quantity of reindeer-moss; and two days later we turned our bows finally northward and eastward, passing through heavy "slack" ice under sail and liquid air in crisp weather, till, on the 27th of August, we lay moored to a floe off the desolate island of Taimur.

The first thing which we saw here was a bear on the shore, watching for young white-fish: and promptly Clark, Mew, and Lamburn (engineer) went on shore in the launch, I and Maitland following in the pram, each party with three dogs.

It was while climbing away inland that Maitland said to me: "When Clark leaves the ship for the dash to the Pole, it is three, not two, of us, after all, that he is going to take with him, making a party of four."

I: "Is that so? Who knows?"

Maitland: "Wilson does. Clark has let it out in conversation with Wilson."

I: "Well, the more the merrier. Who will be the three?"

Maitland: "Wilson is sure to be in it, and there may be Mew, making the third. As to the fourth, I suppose *I* shall get left out in the cold."

I: "More likely I."

Maitland: "Well, the race is between us four: Wilson, Mew, you and I. It is a question of physical fitness combined with special knowledge. You are too lucky a dog to get left out, Jeffson."

I: "Well, what does it matter, so long as the expedition is a success? That's the main thing."

Maitland: "Oh, yes, that's all very fine talk. But isn't it rather a pose to affect to despise $175,000,000? *I* want to be in at the death, and mean to be, if I can."

"Look," I whispered—"a bear."

It was a mother and cub: and with a stubborn trudge she came wagging her low head, having no doubt smelled the dogs. So we separated on the instant, doubling different ways behind ice-boulders, wanting her to go on nearer the shore, before killing; but, in passing close, she spied, and bore down at a trot upon me, whereupon I fired into her neck; and at once, with a roar, she turned tail, making now straight in Maitland's direction. I saw him run out from cover some hundred yards away, aiming his long-gun; but no report followed: and in half a minute he was under her fore-paws, she rapping out slaps at the barking, shrinking dogs. Maitland roared for my help; and at that moment, I, poor wretch, in greater misery than he, stood shivering in an ague: for all at once one of those wrangles of the voices of my destiny was filling my bosom with commotion, one bidding me dash to Maitland's aid, one passionately commanding me be still. But it lasted,

I fancy, some seconds only before I ran and got a shot into the bear's brain; and Maitland leapt up with a rent down his face.

But singular destiny! Whatever I did—if I did evil, if I did good—the result was the same: tragedy dark and sinister! Poor Maitland was doomed that voyage, and my rescue of him was the means employed to make his death the more sure.

I think that I have already written about a man called Scotland, whom I met at Cambridge, who was always taking about certain "Black" and "White" beings, and their contention for the earth; well, with regard to all that, I have a fancy, a whim of the mind, which I will write down now: that there may have been some sort of understanding between Black and White, as in the case of "Adam" and "the tree," that, should mankind force his way to the Pole and the old forbidden mystery biding there, then some mishap should not fail to overtake the race; that the White, being kindly inclined to mankind, did not wish this to take place, and intended, for the sake of the race, to wipe out our entire expedition ere it reached; and that the Black, knowing that the White designed to do this, and by what means, used me—*me*—to outwit this scheme, first of all working that I should be one of the party of four to leave the ship on ski.

But the baby attempt, my God, to read . . . I laugh at poor Black-and-White Scotland! The thing ain't so simple.

Well, we left Taimur the same day, and good-bye now to both land and open sea. Till we passed the latitude of Cape Chelyuskin (which we did not sight), it was one succession of ice-belts, with Mew in the crow's-nest tormenting the electric bell to the engine-room, the anchor hanging ready to drop, and Clark taking soundings. Progress was slow, and the Polar night gathered round us gradually, as we groped still onward and onward into that indigo and glimmering clime of frore, we now leaving off bed-coverings of reindeer-skin to take to sleeping-bags, eight of the dogs having died by the 25th of September, when we were experiencing 19° of frost. In the darkest part of our night the Northern Light cast its solemn gonfalon over us, quivering round the skies in a million fickle gauds.

Meantime, the relations between the members of our little crew were excellent—with one exception: David Wilson and I were not good friends.

There was a something—a tone—in the evidence given by him at the inquest on Peters which made me mad every time I remembered it. He had heard Peters admit that Peters had administered atropine to himself, and had had to give evidence of that fact; but had given it in a most half-hearted way, so much so, that the coroner had asked him, "What, Sir, are you hiding from me?" From which day he and I had hardly exchanged ten sentences, in spite of our constant companionship in the vessel; and one day, standing alone on a floe, I found myself hissing: "If he dared suspect Clodagh of poisoning Peters, I could *kill* him. . . ."

Well, up to 78° of latitude the weather had been splendid, but on the night of the 7th of October—well I remember it—we experienced a fierce tempest. Our tub of a ship rolled like a swing, drenching the whimpering dogs at every lurch, and hurling everything on board into confusion; the petroleum-launch was washed from the davits; down at one time to 40° below zero sank the thermometer; while a high aurora was whiffed into a dishevelled crush of chromes, resembling the palette of some rabid Rafael or mixed battle of seraphim in their robes, and looking the very symbol of tribulation, tempest, wreck, and distraction. I, for the first time, was sick.

It was with a dizzy brain, therefore, that I went off watch to my bunk. Soon, indeed, I fell asleep; but the rolls and shocks of the ship, combined with the ponderous Greenland-anorak which I had on, and the state of my body, together produced a frightful nightmare, in which I was conscious of a vain struggle to move, a vain fight for breath, for the sleeping-bag turned to an iceberg on my bosom. Of Clodagh I dreamt—that she let drip a liquid, colored like pomegranate-seeds, into a glass of gruel: and she presented the glass to Peters. The draught, I knew, was poisonous as death; and in a last effort to break the bands of that dark slumber I was conscious, as I wrenched myself upright, of shrieking aloud: "Clodagh! *spare the man . . . !*"

Now my eyes opened to waking; the electric light was shining in the cabin; and there stood David Wilson looking at me.

Wilson was a big man, with a massively-built face, long, made longer by a beard, having nervous contractions of the flesh at the cheek-bones, and splashed with freckles: I can see him now, his clinging pose, his mouth of disgust, his whole air, as he stood crouching and lurching there.

What he was doing in my cabin I did not know. To think, my good God, that he should have been led there just then! This was one of the four-men starboard berths; *his* was a-port: yet there he was. But he explained at once.

"Sorry to interrupt your innocent dreams," says he: "the mercury in Maitland's thermometer is frozen, and he asked me to hand him his alcohol one from his bunk. . . ."

I did not answer. A hatred was in my heart against this man.

The next day the storm died away, and either three or four days later the slush-ice between the floes froze definitely. The *Boreal's* way being thus blocked, we warped her with ice-anchors and the capstan into the position in which she should lay up for her winter's drift. This was in about 79° 20′ N. The sun had now totally vanished from our bleak abode, not to reappear till the following year.

Well, there was sledging with the dogs, and bear-hunting among the hummocks, as the months, one by one, went by; one day Wilson, by far our best shot, got a walrus-bull; Clark followed the traditional pursuit of a Chief, examining crustacea; Maitland and I were in a relation of close friendship, and I assisted his meteorological observations in a snow-hut built near the ship; sometimes throughout the twenty-four hours a luminous blue moon, very spectral, very fair, imbued our dim and livid dominion.

It was four days before Christmas that Clark made the great announcement; he had decided, he said, if our fine northward drift continued, to leave the ship near the middle of March for the dash to the Pole, taking with him the four reindeer, all the dogs, four sledges, four kayaks, and three companions; the companions whom he had decided to invite being: Wilson, Mew, and Maitland.

He said it at dinner; and, as he said it, David Wilson glanced at my face with a smile of glad malice that *I* was left out.

I remember well: the aurora that night was in the sky, at its brink floating a moon surrounded by a ring, with two mock-moons; but all shone very vague and far, and a fog which had already lasted some days made the ship's bows indistinct to me, as I paced the bridge on my watch, three hours after Clark's announcement.

For a long time all was quite quiet, save for the occasional whine of a dog, I all alone there; and, as it grew toward the end of my watch, when Maitland would follow me, my slow tread tolled as for the grave, the mountainous ice lying vague round me in its shroud and taciturnity, not less dreadfully strange than eternity itself.

But presently several of the dogs began barking together, left off, and began again. I said to myself: "There's a bear about. . . ."

And after some minutes I saw—thought that I saw—it, though the fog had, if anything, thickened: it being now very near the end of my watch.

It had entered the ship, I conjectured, by the boards which slanted from the port gangway down to the ice. Once before, in November, a bear, having smelled the dogs, had ventured on board at midnight; but *then* there had resulted a regular hubbub among the dogs; *now,* even in the midst of my excitement, I wondered at their quietness, though some whimpered—with fear, I thought. I saw the creature steal froward from the hatchway toward the kennels a-port; and I ran noiselessly, to snatch the watch-gun which stood always loaded by the companionway.

By this time the form had passed the kennels, had walked to the bows, was now making toward me on the starboard side; and, as I took aim, never, I thought, had I beheld so immense a bear—though I made allowance for the magnifying effect of the fog.

My finger was on the trigger; and in that instant a shivering sickness took me, the two voices shouting at me, "Shoot!" "Shoot not!" "Shoot!" Ah, well, that latter was irresistible. I pulled the trigger. The report hooted through the Polar glooms.

As the creature dropped, both Wilson and Clark were up at once; and we three hurried to the spot.

But the first near glance discerned a singular species of bear; and when Wilson put his hand to the head a lax skin came away at his touch. . . . It was Aubrey Maitland who was underneath it; and I had shot him dead.

For some days he had been cleaning skins, among them the skin of the bear from which I had saved him at Taimur, and, as Maitland was a born pantomimist, continually inventing hoaxes, perhaps to startle me with a false alarm in the very skin of the creature which had so nearly done for him, he had thrown it round him on finishing its cleaning, then in wanton fun had crept on deck at the hour of his watch; and the head of the bear-skin, and the fog, must have prevented him from seeing me taking aim.

This thing made me ill for many days: for I saw that the hand of fate was upon me. When I rose from bed, poor Maitland was lying in the ice behind the great camel-shaped hummock close by us.

By the end of January we had drifted to 80° 55′; and it was then that Clark, in the presence of Wilson, asked me if I would make the fourth man, in the place of poor Maitland, for the dash in March. When I said "Yes, I am willing," David Wilson spat with a disgusted emphasis; then, a minute later, he sighed, with "Ah, poor Maitland . . ." and drew in his breath, with tut! tut!

God knows, I had an impulse to spring then and there at his gullet, and strangle him; but I restrained myself.

There remained now hardly a month before the dash, and all hands set to work with a will, measuring the dogs, making harness and seal-skin shoes for them, overhauling sledges and kayaks, and cutting out every possible ounce of weight. But we were not destined, after all, to set out that year; about the 20th of February the ice began to pack, subjecting the ship to terrific pressure, while we found it necessary to make trumpets of our hands to shout into each others' ears, the entire ice-continent crashing, popping, crackling on every side in cosmic upheaval; and, expecting every moment to see the *Boreal* cracked to splinters, we had to set about unpacking provisions, and placing sledges, kayaks, dogs and everything in a position for instant flight. Five days it lasted, accompanied by a storm from the north, which, by the end of February, had driven us back south into latitude 79° 40′. Clark, of course, then abandoned all thought of the Pole for that summer.

And immediately afterwards we made a startling discovery: that our stock of reindeer-moss was now somehow ridiculously small. Egan, our second mate, was blamed; but that did not help matters: the sad fact remained; and, since Clark, when begged to kill one or two of the deer, pig-headedly refused, by the beginning of summer every one was dead.

Well, our northward drift recommenced. Toward the middle of February we saw a mirage of the coming sun above the horizon; there were flights of Arctic petrels and snow-buntings; spring was with us; and in an ice-pack of big hummocks and narrow lanes we made good progress all the summer.

When the last of the deer died, my heart had sunk, and when the dogs killed two of their number, and a bear crushed a third, I was expecting what came: Clark announced that he could now take only two companions with him in the spring: Wilson and Mew. So once more I witnessed David Wilson's complacent smile of malice.

Then we settled into our second winter-quarters: again December, and all that moodiness and dreariment of our sunless gloom, made worse by the fact that the wind-mill would not work, leaving us frequently without electricity.

Ah me, none but those who have experienced it could dream one half the mental depression of that Arctic dark; how the soul takes on the hue of the universe; and without and within is nothing but gloom, gloom, and the rule of the Power of Darkness. Not one of us but was in a melancholic, dismal and dire mood; and on the 19th December Lamburn, the engineer, stabbed Cartwright, the old harpooner, in the arm.

Three days before Christmas a bear came close to the ship, then turned tail; upon which Mew, Wilson, I and Meredith (a general hand) set out in pursuit; but after a pretty long chase lost him; then scattered different ways. It was very dim, and after yet an hour's search I was returning tired and dispirited to the ship, when I spied some shade like a bear sailing away on my left, and at the same time sighted a man—I did not know whom—running like a handicapped ghost on my right. So I cried out: "There he is—come on! this way!"

The man quickly joined me, but, as soon as ever he recognized me, stopped dead, and the devil must have suddenly got into him, for he said: "No, thanks, Jeffson: alone with you I am in danger of my life. . . ."

It was Wilson. And I, too, forgetting at once all about the bear, stopped and faced him.

"I see," said I. "But, Wilson, you are going to explain to

Confrontation with Wilson -

me *now* what you mean, you hear? What *do* you mean, Wilson?"

"What I say," he answered deliberately, eyeing me up and down: "alone with you I am in danger of my life: just as poor Maitland was, and just as poor Peters was. Certainly, you are a deadly beast."

Frenzy leapt, my God, in my heart; dark as that darksome Arctic night was my mind.

"Do you mean," said I, "that I want to put you out of the way, in order to go in your place to the Pole? Is that your meaning, man?"

"That's about my meaning, Jeffson," says he: "you are a deadly beast, you know."

"All right!" I cried, with a blazing eye: "I am going to kill *you*, Wilson—as sure as God lives. But I want to hear first: *who* told you that I killed Peters?"

"Your lover killed him—with *your* collusion. Why, I heard you, man, in your beastly sleep, blabbing the whole thing out. And I was pretty sure of it before, only I had no proofs. By God, I should enjoy putting a bullet into you, Jeffson!"

"You wrong me, you—you wrong me!" I bellowed, my eyeballs staring in ravenous lust for his blood; "and now I am going to pay you well for it. *Look out, you!*"

I aimed my gun for his gizzard, I fingered the trigger; but he held up his left hand.

"Stop," he said, "stop." (He was ever one of the coolest of men). "There is no gallows on the *Boreal*, but Clark could easily rig one for you. I want to kill you, too, because there are no criminal courts up here, and it would be doing a good action for my country; but not here—not now: listen to me—don't shoot. Later we can meet, when all is ready, so that no one may be the wiser, and fight it all out."

As he spoke, I let the gun drop: it was better so. I knew that he was much the best shot on the ship, and I an indifferent one: but I did not care, I did not care, if I was killed.

It is a dim, inclement land, God knows; and the spirit of darkness and distraction is there. . . .

Twenty hours later we met behind the great saddle-shaped hummock, some six miles to the S.E. of the ship; had set out at different times, so that no one might suspect; and each brought a ship's-lantern.

Wilson had dug an ice-grave near the hummock, leaving at its edge a heap of brash-ice and snow to fill it; and, this grave between us, we stood separated by perhaps seventy yards, each with his lantern at his feet.

Even so we were just ghosts and shades to each other, the air glowering very drearily, and present in my inmost soul were frills of cold, a chill moon, a mere abstraction of sheen, seeming to hang far outside the universe, the temperature at 54° below zero, so that we had on wind-clothes over our anoraks, and heavy foot-bandages under our Lap-boots. Nothing but a weird morgue seemed the world, haunted with despondent madness; and exactly like that world round us were the bosoms of us two poor men, full of macabre, bleak, and funereal feelings.

Between us yawned an early grave for one or other of our bodies; and I heard Wilson cry out: "Are you ready, Jeffson?"

"Aye, Wilson!" I cried.

"Then, here goes!" cries he.

As he spoke, he fired: surely, the man was in earnest to kill me.

But his shot passed by me, as indeed was only likely, for we were shadows to each other.

I fired perhaps five seconds later than he: but in those five seconds he stood brightly revealed to me in clear lilac light: for an Arctic fireball had shot across the sky, showering abroad a phosphorous shine over the snow-landscape.

Before the intenser blue of its momentary glamour had passed away I saw Wilson stagger forward, and drop. And him and his lantern I buried there under the rubble ice.

* * *

On the 13th March, nearly three months later, Clark, Mew and I left the *Boreal* in latitude 85° 15′.

We had with us thirty-two dogs, three sledges, three kayaks, human provisions for 112 days, and dog provisions for 40. Being now about 340 miles from the Pole, we hoped to reach it in 43 days, then, turning south, and, feeding living dogs with dead, make either Franz Josef Land or Spitzbergen, at which latter place we should very likely come up with a whaler.

Well, during the first days progress was very slow, the ice being rough and laney, and the dogs behaving most badly, stopping dead at every difficulty, and leaping over the traces. Clark had had the idea of attaching a gold-beater's-skin balloon, with a lifting power of 35 pounds, to each sledge, and we had with us a supply of zinc and acid to repair the hydrogen-waste from the bags; but on the third day Mew overfilled and burst his balloon, whereupon Clark and I had to cut ours loose to equalize weights: so at the end of the fourth day out we had made only nineteen miles, and could still from a hummock perceive afar the leaning masts of the old *Boreal*. Clark led on ski, captaining a sledge with 400 lbs. of instruments, ammunitions, pemmican, aleuronate bread; Mew followed, his sledge containing provisions only; and last came I, with a mixed freight. But on the fourth day Clark had an attack of snow-blindness, and Mew took his place.

* * *

Pretty soon our sufferings commenced, and they were bitter

enough: the sun, though constantly visible day and night, gave no heat; our sleeping-bags (Clark and Mew slept together in one, I in another) were soaking wet all the night, being thawed by our warmth; and our fingers, under wrappings of sennegrass and wolf-skin, were always bleeding. Sometimes our frail bamboo-cane kayaks, lying across the sledges, would crash perilously against an ice-ridge—our one hope of reaching land; but the dogs were the great difficulty: we lost six mortal hours a day in harnessing and tending them. On the twelfth day Clark took a single-altitude observation, and found that we were only in latitude 86° 845′; but the next day we passed beyond the farthest point yet (authentically) attained—by the *Nix*.

* * *

Our secret thought now was food, food—our day-long lust for the eating-time. Mew suffered from "Arctic thirst."

* * *

Under such conditions man becomes in a few days, not a savage only, but a brute, scarcely a grade above the bear and walrus. . . . Ah, the ice! A sordid nightmare was that, God knows.

* * *

On we pressed, wending our petty way over the immense, upon whose loneliness, from before the old Silurian till now, Boötes had pored and brooded.

* * *

After the eleventh day our rate of march improved, all lanes disappearing, ridges becoming much less frequent. By the fifteenth day I was leaving behind me the ice-grave of David Wilson at the rate of ten to twelve miles a day.

Yet, as it were, his arm reached out and touched me, even there.

His disappearance had been explained by a hundred different guesses on the ship—all plausible enough: I had no idea that anyone connected me in any way with his death.

But on our twenty-second day of march, 140 miles from our goal, he caused a conflagration of rage and hate to break out among us three.

It was at the end of a march when our stomachs were hollow, our frames ready to drop, and our mood ravenous and inflamed. One of Mew's dogs was sick: it was necessary to kill it; he asked me to do it.

"Oh," I said, "you kill your own dog, of course."

"Well, I don't know," he replied, catching fire at once, "you ought to be used to killing, Jeffson."

"How do you mean, Mew?" I asked, with a mad start, for

madness and the lamps of Hell were prompt and ready in us all: "you mean because my profession——"

"Profession, damn it, no," he snarled like a dog: "go and dig up David Wilson—I dare say you know where to find him—*he*'ll tell you my meaning, right enough."

I rushed at once to Clark, who was stooping among the dogs unharnessing, and, savagely pushing his shoulder, I exclaimed: "That beast accuses me of murdering David Wilson!"

"Well?" says Clark.

"I'd split his skull as clean——!"

"Go away, Adam Jeffson, and let me be!" Clark snarled.

"Is that all you've got to say about it, then, you?" I asked.

"To the devil with you, man, say I, and let me be!" he cried: "*you know your own conscience best*, I suppose."

Before this insult I stood with grinning teeth, but impotent, though from that moment a still grummer mood of malignity brooded in my spirit; and indeed the humor of each of us three was imbued with a certain dangerous, even murderous, rage: for in that region of chill we had become assimilated to the beasts that perish.

* * *

On the 10th of April we passed the 89th parallel, and, though sick to death, both in spirit and body, pressed still on. Like the lower animals we were smitten now with dumbness, and hardly once a day mumbled a syllable one to the other; but in selfish brutishness on through a hell of cold we moved. It is damned territory, not to be penetrated by man: and rapid and deplorable was the degeneration of our souls. As for me, never could I have imagined that savagery so heinous could brood in a human bosom as now I felt it brood in mine. If men could enter a country specially set apart for the habitation of devils, and there become possessed of evil, as we were so would they be.

* * *

As we advanced, the ice every day became smoother: so that, from four miles a day, our rate increased to fifteen, and finally (as the sledges lightened) to twenty.

It was now that we began to encounter a succession of strange-looking objects lying scattered across the ice, whose number continually increased as we proceeded, objects having the appearance of rocks, or pieces of iron-ore, incrusted with glass-like fragments, which we discovered to be precious stones. On our second twenty-mile day Clark picked up a diamond-splinter as large as a child's thumb, and such objects became common. We thus found "wealth," beyond dream; but as the bear and the walrus find, and for all those millions we would not have given an ounce of fish-meal. Clark grumbled something about their being meteor-stones, whose ferruginous substance had been lured that way by the Pole's magnetism, and kept from frictional ignition in their passage through the air by the frigidity there: but, as the Pole's H is not strong, my own view is that they are due to the greater drag of gravity and the much greater shallowness of the atmosphere there; anyway, they quickly ceased to interest our sluggish brains, except in so far as they obstructed our way.

* * *

We had all along had excellent weather, till, on the morning of the 12th of April, we were overtaken by a storm from the S.W. of such monstrous and solemn volume, that the heart quailed under it. It lasted in its full power only an hour, but during that time snatched two of our sledges far away, and compelled us to lie face-downward. As we had travelled all the sun-lit night, we were gasping with fatigue: so, as soon as the wind allowed us to huddle together our scattered things, we collapsed into the sleeping-bags, and instantly slept.

We knew that the ice was in fearful upheaval round us; we knew, as our eyelids sweetly closed, of a slow booming as of distant cannon, and brittle cracklings of musketry. This may have been a result of the tempest rumpling-up the sea beneath the ice; whatever it was, we did not care: we slept deep.

We were within nine miles of the Pole.

* * *

In my dream it was as though some messenger shook my shoulder with an urgent "Up! Up!"; nor was it either Clark or Mew, for Clark and Mew, when I started up, lay there in their sleeping-bag.

I suppose it must have been about noon. There I sat staring some minutes, and my numb memory was of this: that the Countess Clodagh had prayed me "Be first"—for her. Wondrous little cared I now for the Countess Clodagh in her unreal world of warmth, wondrous little for the fortune which she coveted: fortunes swarmed unregarded on the ground round me; yet that urging, *"Be first!"*, was profoundly suggested in my spirit, as if whispered within my inwards: and instinctively, brutishly, as the Gadarean swine rushed down a steep place, I, rubbing my daft eyes, arose.

The first thing which my mind opened to note was that, while the tempest was less strong, the ice was at present in extraordinary agitation, I looking abroad upon a plain stretched out to a waving horizon, varied by hillocks, boulders, and glimmering meteor-stones that everywhere tinselled the blinding white, some big as wire-guns, most little as limbs; and this vast plain was at present rearranging itself in a far-spread drama of havoc, withdrawing in chasms like mutual backing curtsies, then surging to clap together in passionate mountainpeaks, else jostling like the Symplegades, nimbly inconstant as billows of the sea, grinding itself, piling itself, pouring itself in downfalls of powdered ice, while here and there I saw the meteor-stones leap spasmodically, in dusts and heaps, like geysers or hopping froths in a steamer's wake, all the trumpets of uproar, meanwhile, occupying the air. In standing, I tripped and staggered, and saw all the dogs sprawling, with whimperings of misery.

I did not care. Instinctively, daftly, brutishly, I harnessed ten of them to my sledge; put on Canadian snow-shoes: and was away northward—alone.

The sun shone with a clear, benign, but heatless shining, a ghostly, remote, yet limpid light, which seemed designed for the lighting of other planets and systems, and to strike here by happy chance. A wild wind from the S.W., meanwhile, flung thin snow-sweepings flying northward past me.

My odometer had not yet measured four miles, when I commenced to note two things: one that the meteor-stones were now accumulating beyond limit, filling my field of vision to the northern horizon with a blinding brightness, lying in piles, in parterres, like largesse of autumn leaves, so that I had need to steer my feet among them; now, too, I noticed that, but for these stones, all roughness had disappeared, not a trace of the upheaval going on a few miles south being here: for the ice lay nearly as smooth as a table before me, and it is my belief that this stretch of smooth ice has never, never, felt shock or throe, but reaches right down to the bottom of the deep.

* * *

And now with a wild hilarity I flew, for a lunacy, a giddiness, had got me, until finally, up-buoyed on air, dancing mad, I sped, I span, with grinning teeth that chattered and gibbered, and eyeballs of distraction: for a fright, too—most cold, most mighty high—had its hand of ice on my soul, I being alone in that place, face to face with the Ineffable; but still, with a gibbering levity, and a fatal joy, and a blind hilarity, on I sped, I span.

* * *

The odometer measured nine miles from my start: I was in the neighborhood of the Pole.

I cannot say when it began, but now I was conscious of a sound in my ears, clear and near, a steady sound of splashing, or fluttering, resembling the noising of a cascade or brook; and it grew. Forty more steps I took (skate I could not now for the meteorites)—perhaps eighty—perhaps a hundred: and now, to my sudden horror, I stood looking at a lake.

One minute, swaying and nodding there, I stood, then dropped down flat in swoon.

* * *

In a hundred years, I suppose, I should never succeed in analyzing *why* I swooned: but my consciousness still retains the impression of that horrid thrill. I saw nothing distinctly, for my being reeled and toppled drunken, like a spinning-top in desperate death-struggle at the instant when it flags, and wobbles dissolutely to fall; but the moment my eyes lighted on what lay before me—a lake, circular, clean-cut—I felt, I fathomed, that here was the sanctuary, here the eternal secret of this earth from her birth, which it was a burning shame for a worm to see. The lake, I think, would be something like a mile wide, and in its middle is a pillar of ice, low and thick; and I had the impression, or dream, or fantasy, that there is a name inscribed round in the ice of the pillar in characters that could never be read; and under the name a lengthy date; and the liquid of the lake seemed to me to be wheeling with a shivering ecstasy, splashing and fluttering, round the pillar, from west to east, with the planet's spin; and it was borne in upon me—can't say how—that this fluid was the substance of a living being; and I had the fancy, as my senses failed, that it was a being with many eyes, dull, repining, and that, as it swept for ever round in fluttering lust, it kept its many gazes riveted on the name and the date graven in the pillar. But some of this must be my madness. . . .

* * *

It must have been not less than an hour before a sense of life arose again in me; and when the thought broke in upon my brain that a long, long time I had lain there in the presence of those gloomy orbs, my spirit groaned and died within me.

In some minutes, however, I had scrambled on my legs, caught at a dog's harness, and without one backward glance was escaping from that place.

Half-way to the halting-place I awaited Clark and Mew, being very sick and doddering, and unable to advance. But they did not come.

Later on, when I gathered force to go farther, I found that they had perished in the upheaval of the ground. One only of the sledges, half buried, I saw near the spot of our bivouac.

* * *

Alone that same day I began my way southward, and for four days made good progress. On the seventh day I noticed, stretched right across the south-eastern horizon, a region of vapor which luridly obscured the face of the sun; purple it looked, and day after day I observed it steadily brooding there; but what it could be I did not know.

* * *

Well, onward through the desert I went my solitary way,

with a quailing terror in me: for very stupendous, alas, is the load of that Polar lonesomeness on one poor human soul.

Often on a halt I have lain and listened long to the hollow stillness, recoiling, appalled by it, longing that at least one of the dogs might whimper; I have even crawled quivering from the thawed sleeping-bag to flog a dog, so that I might hear a voice.

* * *

I had started from the Pole with a well-filled sledge, and with the sixteen dogs left alive from the ice-packing which had engulfed my comrades, having saved from the wreck of our things most of the whey-powder, pemmican, &c., as well as the theodolite, compass, chronometer, train-oil lamp for cooking, and other implements: I was therefore in no doubt as to my course, and had provisions for eighty days; but ten days from the start my stock of dog-food failed: I had to begin to slaughter my companions, one by one; and in the third week, when the ice became horribly rough, with enough moil and toil to wear a bear to death I did only five miles a day. After the day's work I would creep with a dying sigh into the sleeping-bag, clothed still in the load of skins which stuck to me a mere filth of grease, to sleep the sleep of a pig, indifferent if I never woke.

And ever—day after day—about the south-eastern heaven brooded heavily that curious region of purple vapor, streaming like the smoke of the conflagration of the world, its length steadily growing.

* * *

Once I had a pretty agreeable dream—dreamed that I was in a garden—an Arab paradise—sweet to breathe; yet—all the time—I had a sub-consciousness of the storm which was actually blowing from the S.E. over the ice, and at the moment when I awoke was half-wittedly mumbling to myself: "It is a garden of peaches; but I am not really in the garden: I am really in the Arctic; only, the S.E. gusts are wafting to me the aroma of this garden of peaches."

I opened my eyes—I started—I sprang to my feet! For, mad as it was, I could not doubt—an actual aroma like peach-blossom *was* in the algid air about me!

Before I could collect my astonished senses I began to vomit violently, and at the same time saw some of the dogs, skeletons as they were, vomiting also; then for a long time I lay sick in a kind of daze; and, on getting up, found three of the dogs dead, and all very queer. The wind had now changed to the north.

Well, on I stumbled, fighting each inch of my deplorably weary way, this odor of peach-blossom, my sickness, and the death of the three dogs, remaining a wonder to me.

Two days later I came across a bear and her cub lying dead at the foot of a hummock, and could not believe my eyes: there she lay, a spot of dirty-white in a disordered patch of snow, with one little eye open, and her fierce-looking mouth also; and the cub lay across her haunch, biting into her rough fur. So I set to work upon her, and allowed the dogs a glorious feed on the blubber, while I myself had a banquet on the fresh meat; but then had to leave the greater part of the carcasses, and I can feel again now the hankering reluctance with which I trudged onwards. Again and again I found myself asking: "Now, what could have killed those two bears?"

With brutish stolidness I plodded ever on, almost like a walking machine, sometimes nodding in sleep, while I helped the dogs, or maneuvered the sledge over an ice-ridge, pushing or pulling. On the 3rd of June, a month and a half from my start, I took an observation with the theodolite, and found that I was not yet 400 miles from the Pole, in latitude 84° 50′. It was as though some will was obstructing me.

However, the intolerable cold was over, and soon my clothes no longer hung stark on me like armor; pools began to appear in the ice, and presently, what was worse, my God, long lanes, across which, somehow, I had to get the sledge. But about the same time all fear of starvation passed away: for on the 6th of June I came across another dead bear, on the 7th three, and thenceforth, in rapidly growing numbers, I met, not bears only, but fulmars, guillemots, snipes, Ross's gulls, little awks—all, all, lying dead on the ice, never anywhere a living thing, save me, and the two remaining dogs; and if ever a poor man stood shocked before a mystery, it was I now.

On the 2nd of July the ice began packing dangerously, and soon another storm broke loose upon me from the S.W.: so I left off my trek; put up the silk tent on a five-acre square of ice surrounded by lanes; and it was there that *again*—for the second time—as I lay down, I smelled that delightful strange odor of peach-blossom, a mere whiff, and presently was taken sick. However, it passed off this time in half an hour.

Now it was all lanes, lanes, alas, yet not open water, and such was the drudgery and woe of my life, that sometimes I would drop prostrate upon the ice, sobbing "Oh, no more, my God, here let me die." The crossing of a lane might occupy ten, twelve hours, and then, on the other side, I might find another one opening right before me. Moreover, on the 9th of July, one of the dogs, after a feed on blubber, suddenly died, leaving me only "Reinhardt," a white-haired Siberian dog, with little brisk up-sticking ears, like a cat's; and him also I had to kill on coming to open water.

This did not happen till the 3rd of August, nearly four months from the Pole.

I can't think, my God, that any soul of man ever tholed that dismal incubus or that abysm of sensations within which, during those four months, I weltered: for, though I was as a brute, I had a man's heart to smart. What I had seen, or dreamed, at the Pole followed and followed me; and, if I shut my eyes to sleep, those other eyes yonder seemed to watch me again with their distraught and gloomy gaze, and in my dark dreams reeled that everlasting ecstasy of the lake.

However, by the 28th of July I knew from the look of the sky, and the absence of fresh-water ice, that the sea could not be far: so I set to business, and spent two days in putting to rights the now battered kayak. This done, I had no sooner resumed my way than I sighted on the horizon a streaky haze, which could only be the cliffs of Franz Josef Land; and in a craziness of jubilation I stood there, waving my ski-staff round my head, with the senile cheers of an aged man.

In three days this land was visibly nearer, sheer basaltic cliff mixed with glacier, forming apparently a great bay, with three islands in the mid-distance! and at dawn of the 5th of August I arrived at the definite limit of the pack-ice in moderate weather near the freezing-point.

At once, but with great reluctance, I shot Reinhardt, then set to getting the last of the provisions, and the most necessary of the implements, into the kayak, making haste to put out to the luxury of being borne on water after all the trudge; and within fourteen hours was coasting, with my little lug-sail bulged, along the shore-ice of that land: the midnight of a calm Sabbath; and low down on the horizon smoked the ruddy sunball drowsing, as my canvas skiff lightly chipped her passage through that silent sea. Silent, silent: for neither snort of walrus, nor wawl of fox, nor screech of kittiwake, did I hear: but all was still as the jet-black shadow of cliff and glacier on the sea; and many corpses of dead things swarmed on the face of the water.

* * *

When I found a fjord I wound up it to the end, where stood a stretch of basalt columns, looking like a shattered temple of Antediluvians; and when my foot at last touched land, I dropped there bowed down a long, long while in the rubbly snow, and silently wept, my eyes that night a fountain of tears: for the firm land is health and sanity, and dear to the life of man, but the ice is a nightmare, and a blasphemy, and a madness, and the realm of the Power of Darkness.

* * *

I knew that I was at Franz Josef Land somewhere in the neighborhood of C. Fligley (about 82° N.); and, though it was so late, and getting cold, I still had the hope of reaching Spitzbergen that year, by alternately navigating the open sea

Winter shelter

and dragging the kayak over the slack drift-ice. As all the ice which I saw was good flat fjord-ice, the plan appeared feasible enough; so, after coasting about a little, and then three days' rest in the tent at the bottom of a ravine of columnar basalt opening upon the shore, I packed some bear and walrus flesh, with what artificial food was left, into the kayak, and set out in the morning, coasting the shore-ice with sail and paddle until the afternoon. Then, on managing to climb a little way up an iceberg, I made out that I was in a bay whose terminating headlands were invisible: so I determined to make straight S.W. by W. to cross it; but, in doing so, I was hardly out of sight of land when a northern storm overtook me toward midnight, and, before I could think, the little sail was all but whiffed away, the kayak upset. I only saved it by the happy chance of being near a floe with an ice-foot, which, jutting out under the waves, gave me foot-hold; and on the floe I lay in a mooning state the whole night through under the tempest's piping, for I was half drowned.

Happily, my instruments, etc., had been saved by the kayak-deck when she capsized; but I now abondoned all thought of whalers and of Europe for that year.

* * *

A hundred yards inland from the shore-rim, in a place where there was some moss and soil, I built myself a semi-subterranean Eskimo-den for the Polar night, the spot surrounded by high walls of basalt, except to the west, where they opened in a cleft to the coast, the ground strewn with slabs and boulders of granite and basalt, in three places the snow red, overgrown with a lichen which at first I took for blood; and I found in there a dead she-bear, two cubs, and a fox, the last fallen from the cliffs; but I did not even yet feel secure from possible bears, and took care to make my den fairly tight, a job which occupied me nearly four weeks: for I had no tools, save a hatchet, knife, and metal-shod ski-staff. I dug a passage in the ground two feet wide, two deep, ten long, with perpendicular sides, and at its north end I dug a round space, twelve feet across, with perpendicular sides, which I lined with rocks; the whole excavation I covered with walrus-hide, inch-thick, skinned during a bitter week from four of a number which lay about the shore-ice; and for ridge-pole I used a rock-splinter which I found, though, even so, the roof remained nearly flat. This, when finished, I stocked well, putting in everything, except the kayak, blubber both for fuel and occasional light, and foods of several kinds, procured by just stretching out the hand. The roof of both round part and passage was soon buried under snow, and hardly distinguishable from the general level of the ground; and through the passage, if I passed in or out, I prowled on hands and knees; but that was seldom: and within the little round interior, mostly seated, cowering, with quiverings, I wintered, hearkening to the mouthings of darkling storms that bawled about my forlornness.

* * *

All those months the burden of a thought bowed me, and a question like the slow turning of a mechanism worked in my melancholy soul: for everywhere round me lay bears, walruses, foxes, thousands upon thousands of little awks, kittiwakes, snow-owls, eider-ducks, gulls—dead; almost the only living things which I saw being some walruses on the drift-floes, but very few of these: and it was clear to me that some inconceivable catastrophe had overtaken the island during the summer, destroying all life about it, except some few of the amphibia, cetacea, and crustacea.

On the 7th of December, having crept out from the den during a southern tempest, I had, for the third time, a distinct whiff of that self-same smell of peach-blossom; but now without any after-effects.

* * *

Well, again came Christmas, the New Year—Spring: and on the 22nd of May I set out with a well-stocked kayak, the water now fairly open, and the ice so good, that at one place I could sail the kayak over it, the wind sending me sliding at a fine pace. Being on the west coast of Franz Josef Land, I was in as favorable a situation as could be, and I bent by bow southward with much hope, keeping for days just in sight of land; but toward nightfall of my fourth day out, on noticing a floe that presented a lovely sight, looking freighted with a profusion of roses which it reflected within its crystal, I went to it, and saw it covered thick with millions of Ross's gulls, all dead, whose rosy bosoms had given it that bloom.

Well, up to the 29th of June I made good progress southward and westward, the weather mostly excellent, I sometimes coming on dead bears floating away on floes, sometimes on dead or living walrus-herds, with troop after troop of dead kittiwakes, glaucus and ivory gulls, skuas, every kind of Arctic fowl; and on that last day—the 29th—as I was about to encamp on a floe soon after midnight, happening to look toward the sun, my eye fell upon something far away south across the ocean of floes—*the masts of a ship.*

A phantom ship, or a real ship: it was all one to me; real, I must instantly have felt, it could hardly be; but at a sight so wild my heart set to beating as though I must die, and feebly waving the cane oar about my head, I collapsed upon my knees, and thence toppled flat.

So overpoweringly sweet was the prospect of springing once more, like the beasts of Circe, from a walrus into a European: for at this time I was tearing my bear's-meat just

AJ comes upon The Boreal

like a bear, was washing my hands in walrus-blood, to give them a glairy sort of pink cleanness in place of the inky grease that chronically smeared them.

And, worn as I was, I made little delay to set out for that ship; nor had I travelled over water and ice four hours when, to my indescribable joy, I made out from the top of a tallish floe that she was the *Boreal*.

It seemed most strange that she should be anywhere hereabouts! I could only conclude that she must have forced and drifted her way thus far westward out of the ice-block in which our party had left her, and perhaps now was loitering here in the hope of picking us up on our way to Spitzbergen.

In any case, crazy was the rage with which I fought my way to be at her, my gasping lips all the time drawn back in a rictus of laughter at the anticipation of their gladness to see me, of their excitement on hearing the grand tidings of the Pole attained, I anon waving the paddle, although I knew that they could not yet spy me, and then I lashed wildly at the whiteish water. What astonished me was her mainsail and foremast squaresail—set that calm morning, her screws still, for she moved not at all there under a sun which was abroad like a cold spirit of light, touching the ocean-room of floes with blinding spots, a tint almost of rose touching all things, as it were of a just-dead bride in her brilliants and white array, the *Boreal* the one little ink-black spot in all this purity: and upon her, as though she were paradise, I paddled, I panted.

But she was in a queerish state: by 9 a.m. I could see that: two of the windmill-arms not there, and, half-lowered down her starboard beam, a boat hanging askew; moreover, soon after 10, I could see that her mainsail had a rent down the middle. And I could not at all make her out: she was not anchored, though a sheet-anchor was hanging at the starboard cathead; she was not moored; and two small ice-floes, one on each side, were idly bombarding her bows.

I began now to wave the paddle again, battling for my breath, ecstatic, crazy with excitement, each second like a year to me; and when I could now make out someone at the bows, bending well over, looking my way, and something put it into my head that it was Sallit, I set to mouthing an impassioned shouting of "Hi! Sallit! Hallo! Hi!"

I did not see him move, but there he stood, leaning steadily over, looking my way, between me and the ship now being all navigable sea among the ice-floes, and the sight of him so visibly near put into me such a shivering of eagerness, that I was nothing less than demented for the time, sending the kayak flying with venomous digs in sprints, mixing with the diggings my crazy wavings, and with both a hullabaloo of bellowings: "Hallo! Hi! Bravo! *I have been to the Pole!*"

Well, vanity, vanity. Nearer still I drew: broad morning now, going on toward noon, I half a mile away, fifty yards;

but on board the *Boreal*, though now they *must* have heard me, seen me, I observed no movement of welcome, but all, all was still as death that still Arctic morning, my God; only, the ragged canvas flapped languidly, and, one on each side, two ice-floes sluggishly bombarded the bows, with dumb sounds.

I was sure now that Sallit it was who looked across the sea, but when the ship swung a little round, I noticed that the direction of his gaze was carried with her movement, he no longer looking my way; and, "Why, Sallit!" I shouted with reproach at him: "why, Sallit, man!" I whined.

But, even as I shouted and whined, a perfect wild certainty was in me: for a perfume like peach, my God, had now been whiffed from the ship upon me, and I must have very well known then that that watchful outlook of Sallit saw nothing, but on the *Boreal* were dead men all; in fact, I soon saw one of his eyes looking like a glass eye when it slides awry and glares all distraught; and then again my body failed, and my head dropped forward, where I sat, upon the kayak's deck.

* * *

Well, after a long while I started up to look anew at that forlorn and wandering craft: there she lay, quiet, tragic, as it were culpable of the dark cargo of fatality which she bore; there stared Sallit: and I knew quite well why he was there—had leant over to vomit, and had leant ever since, his forearms propped upon the bulwark-beam, his left knee pressing on the boards, his left shoulder propped upon the cathead, his face shaking in response to every bump of the two floes upon the bows, nodding a little, he, strange to say, having no covering on his head, and I noted the play of the zephyrs in his uncut hair. Now I would approach no more, for I was afraid, I did not dare, the stillness of the ship was so sacred; and until late afternoon I sat there watching the black bulk of her hull, watching above her water-line a half-floating fringe of seaweed, proving old sleepiness. An attempt had apparently been made to lower, or take in, the larch-wood pram, for there she hung by a jammed davit-rope, stern up, bow in water; the only two arms of the windmill were moving this way and that, through some three degrees, creaking with an *andante* singsong; some clothes, tied on the bow-sprit rigging to dry, were still there; the iron casing round the bluff bows now red and rough with rust; at several places the rigging in considerable tangle; the boom occasionally moving through the sector of a circle with a tormented skirling cadence; and the sail, rotten, I suppose, from exposure—for she had certainly encountered no heavy weather—gave out anon a ponderous languid flap at a rent down the center. Except Sallit, looking out there where he had jammed himself, I saw no one.

A ship of dead men

By a paddle-stroke now, and another presently, I had closely approached her about four in the afternoon, though my awe of the vessel was complicated by that perfume of hers, whose baleful effects I knew. My tentative approach, however, proved to me, when I remained unaffected, that, here and now, whatever danger there had been was past; and at last, by a hanging rope, with a thumping desperation of heart, I clambered up her beam.

* * *

They had died, it seemed, quite suddenly, for almost all the twelve were in attitudes of activity: Egan in the very act of ascending the companion-way, Lamburn sitting against the chart-room door, apparently cleaning two carbines, Odling at the bottom of the engine-room stair seemed to be drawing on a pair of reindeer komagar, and Cartwright, who was often in liquor, had his arms frozen tight round the neck of Martin, whom he seemed to be kissing, they two lying stark at the foot of the mizzen-mast.

Over all—over men, decks, rope-coils—in the cabin, in the engine-room—between skylight leaves—on every shelf, in each cranny, lay an ash or dust, impalpably fine, purplish; and, steadily reigning through the ship, like the very spirit of death, that perfume of peach.

* * *

Here it had reigned, as I could see from the log-dates, from the rust on the machinery, from the look of the bodies, from a hundred indications, during something over a year: it was, therefore, mainly by the wayward workings of winds and currents that this mortal ship had been brought hither to me.

And this was the first overt intimation which I had that Power (whoever and whatever It or They may be), which through history had been so very careful to conceal Its Hand from men, hardly any longer intended to be at the pains to conceal Its Hand from *me:* for it was just as though the *Boreal* had been openly presented to me by an Agency which, though I could not see, I could readily apprehend.

* * *

The dust, though quite thin and flighty above-decks, was lying thickly deposited below; and, after having made a tour of investigation, the first thing which I did was to examine that—though I had tasted nothing all day, and was exhausted to death. I found my own microscope where I had left it in the box in my berth to starboard, though I had to lift up Egan to get at it, and to step over Lamburn to enter the chart-room; but in there, toward evening, I sat at the table and bent to see if I could make anything of the dust, while it seemed to me as if the myriad spirits of men that have so-

AJ examines the purple dust

journed on the earth, and angel and devil, and Time and Eternity, hung silent round for my verdict: and such an ague had me, that for a long while my wandering finger-tips, all ataxic with agitation, eluded every delicate effort which I made, and I could nothing do.

Of course, I knew that an odor of peach-blossom, resulting in death, could only be associated with some effluvium of cyanogen, or of hydrocyanic ("prussic") acid, or of both: so when I at last managed to examine some of the dust I was not surprised to find among the mass of ash some yellow crystals which could only be potassic ferrocyanide. What potassic ferrocyanide was doing on board the *Boreal* I did not know, nor had I either the means, or the force of mind, to dive then deeper into it; I understood only that by some means the air of the region just south of the Polar environ had been impregnated with a gas which was either cyanogen, or some product of cyanogen; also, that this gas, which is very soluble, had by now either been dissolved by the sea, or else dispersed into space, leaving its faint perfume; and, seeing this, I let my abandoned head drop upon the table, and long I sat there staring crazy: for I had a suspicion, my God, and a fear, in me.

* * *

The *Boreal,* I found, contained sufficient provisions, untouched by the dust, in cases, casks, &c., to last me, probably, forty years: for after two days, when I had scrubbed and boiled some of the filth of fifteen months from my skin, and solaced myself with better food, I overhauled her thoroughly; then spent three more days in oiling and cleaning the engine; then, all being ready, dragged my twelve dead and laid them together in two rows on the chart-room floor; which done, I hoisted for love the poor little kayak which had served me through so many tribulations; and at nine in the morning of the 6th of July, a week from my first sighting of the *Boreal,* I descended to the engine-room to set out.

The screws, in the modern way, were driven by a stream of liquid air exploding through capillary tubes into slide-valve chests, a motor which gave her, in spite of her bluff bulk, sixteen knots; and it is the simplest thing for one to take these crafts round the globe, since their starting depends upon nothing but the depressing of a lever, provided that one does not get blown to the sky, as liquid air, in spite of its ten blessings, does blow people. At any rate, I had tanks of air to last me through twelve years' voyaging, and there was the machine for making it, with forty tons of coal, in case of need, in the bunkers, and the two Belleville boilers, so that I was well off for motors.

The ice, too, was quite slack here, and I do not believe I ever saw Arctic weather so bright and blithe, the temperature at 41°. I found that I was midway between Franz Josef and Spitzbergen, in lat. 79° 23′, long. 39°; my way was clear; and something like a mournful hopefulness was in me, as the engines slid into their rhythmic turmoil, and those screws started to churn the Arctic sea, while I, darting up, took my stand at the wheel: and the bows of my bark bent southward and westward.

* * *

When I needed food or sleep, the vessel slept, too; then went on her way anew.

Sixteen hours a day sometimes I stood sentinel at the wheel, overlooking the varied sameness of the ice-sea, until my knees would give, some delicate steering being frequently required among the floes and bergs, I by now, however, less burdened with my ball of Polar clothes, standing almost slim in a Lap great-coat, a round Siberian fur-cap on my head.

At midnight when I flung myself into my old berth, it was just as though the engines, subsided now into silence, were a dead thing, and had a ghost that haunted me, for I heard them still, and yet not them, but the silence of their ghost; and often I would startle from sleep, horrified to the heart at some sound of exploding iceberg, or bumping floe, noising far through that white mystery of quietude, within which the floes and bergs were like floating tombs, the world a liquid churchyard; nor ever could I be able to express the strange Doomsday shock with which such a booming would recall me from depths of chaos to recollection of myself: for oftentimes, both waking and in nightmare, I did not know on which orb I was, nor in which age, but felt my being adrift in the great gulf of space and eternity and circumstance, with no bottom for my consciousness to stand upon, the world all mirage and a strange show to me, and the frontiers of dream and waking lost.

Well, the weather was most fair all the time, and the sea like a pond. During the morning of the fifth day, the 11th of July, I entered, and went moving down, an extraordinary long avenue of snowbergs and floes, most regularly placed, half a mile perhaps across and miles long, like a Titanic double-procession of statues, or the Ming Tombs, but mounting and sinking as to music on the swell, some towering high, throwing placid shadows on the aisle between, many being of a pellucid emerald hue, three or four pouring down waterfalls which wawled a far and chaunting sound, the sea of a singular thickness, almost like egg-white, while, as always there, some snow-clouds, white and woolly, floated in the pale sky: and down this aisle, which produced a mysterious impression of Cyclopean cathedrals and queer sequesteredness, I had hardly passed a mile, when I sighted a black object at its end.

I rushed to the shrouds, soon made out a whaler: and anew the same panting agitations, rage to be at her, at once possessed me, as I flew to the indicator, put the lever at full, then back to give the wheel a spin, then up the mainmast ratlins, waving a foot-bandage of vadmel tweed snatched up at random; and by the time I was within five hundred yards of her had lashed myself to such a pitch of passion, that I was anew shouting that futile lunacy: "Hullo! Hi! Bravo! *I have been to the Pole!*"; and those twelve dead that I had there in the chart-room must have heard me, and the men on the whaler must have heard me, and smiled their smile.

For, as to that whaler, I should have known better at once, if I had not been doting, since she *looked* like a ship of death, her boom slamming to port or to starboard on the heave of the sea, her foresail reefed that serene forenoon; but only when I was almost upon her, and was rushing down to stop the engine, did the real truth suddenly drench my heated brain; and I nearly ran into her, I was so stunned.

Later I lowered the kayak, and boarded her. . . .

This ship had been stricken into stillness in the thick of a briskness of activity, for I saw not one of her sixty-two who had not been busy, except one boy—a thing of 600 tons, ship-rigged, with an auxiliary engine, armor-plated about the bows; and there was hardly any part of her which I did not overhaul. They had had a great time with whales, for a great carcass, attached to the ship's side by cant-purchase tackle, had been in process of flensing and cutting-in, and on the deck were two blankets of blubber, looking a ton-weight, surrounded by twenty-seven men in many attitudes, some terrifying, some disgusting, several grotesque, the whale dead, and the men dead, too, and death was there, and the germs of nonentity flourishing, and a mesmerism, and a dumbness, whose realm was confirmed, and its government growing old. Four of them who had been removing the gums from a mass of stratified whalebone at the mizzen-mast foot were quite imbedded in whale-flesh; also, in a barrel lashed to the main topgallant mast-head was visible the head of a man with a long-pointed beard, looking out over the sea to the S.W., which made me notice that five only of the probable eight or nine boats were on board; and after visiting the 'tween-decks, where I saw quantities of stowed whalebone-plates, and fifty or sixty oil-tanks, and cut-up blubber; and after visiting cabin, engine-room, fo'cas'le, where I saw a lonely boy of fourteen whose hand was grasping a bottle of rum under the clothes in a box, he at the moment of death being intent upon concealing it—after two hours' search of the ship I returned to my own and started again, to come half an hour later upon all the three missing whale-boats about a mile apart: so I steered zigzag near, to find in each five men and a steerer, and one had the harpoon-gun fired, with the line coiled round and round the chest of the stroke line-manager; and in the others hundreds of fathoms of coiled rope, with toggle-irons, whale-lances, hand-harpoons, and dropped heads, and grins, and lazy *abandon,* and eyes that glared, and eyes that dozed, and eyes that winked.

* * *

After this I began to sight ships not infrequently, and used regularly to have the three lights going at night. On the 12th of July I met one, on the 15th two, on the 16th one, on the 17th three, on the 18th two—all Greenlanders, I think: but of the nine I boarded only three, the glass revealing from afar that on the others was no life; and on the three dead men: so that that suspicion which I had, and that fear, grew heavy upon me.

I went on southward, day after day, sentinel there at the wheel: clear sunshine; the sea sometimes seeming mixed with regions of milk by day; and at night the immense desolation of a globe glimmered-on by a sun ages ago dead, and by a light that was gloom. It was like Night white in death then; and wan as the very realm of death and Hades I have beheld it, most terrifying, that neuter state and limbo of nothingness, when unreal sea and spectral vault, all confines lost, mingled in a void of ghostly phantasmagoria, pale to huelessness, at whose center I, as it were annihilated, seemed to moon aswoon in immensity of space; into which disembodied world would be flirted anon whiffs of that perfume of peach which I knew, and their frequency grew; but onward the *Boreal* moved, traversing, as it were, bottomless Eternity, and I got to latitude 72°, not far now from Northern Europe.

And now, as to that peach-scent—even while some floes were yet round me—I was just like some fantastic mariner, who, having set out to seek for Eden and the Blessed Islands, finds them, and balmy gales from their gardens stream out, while he is yet afar, to greet him with their fragrances of almond and cornel, champac and jasmin and lotus: for, I having now reached a zone where the peach-aroma was constant, all the world seemed embalmed in its perfume, and I could imagine my bark to be journeying further than the earth's verge toward some clime of eternal spice and delightsomeness.

* * *

Well, I saw at last what whalers used to call "the blink of the ice"—its bright apparition or reflection in the sky when left behind, or not yet arrived at, by which time I was in a region where many craft of various kinds were to be seen; I was continually meeting them; and not one did I omit to investigate, while many I boarded in the kayak or the larch-wood pram. Just below lat. 70° I came upon a fleet of what I believed to be Lafoden cod-and-herring fishers, which must have drifted somewhat on a northward current, all loaded with curing fish, and I cruised from one to the other on a zigzag course, they being widely scattered, some mere sand-grains to the glass on the horizon, the evening still and clear with that astral Arctic clarity, the sun just reclining to his low-couched nightly drowse. These brown boats stood rocking there with slow-creaking noises, as of creatures whining in sleep, quite unharmed so far, awaiting the gales of the winter's drama of wrath on that gloomy sea, when a darksome doom, and a deep grave, would not fail them. The fishers were braw carles, having fringes of beard well back from the chin-point, and hanging woolen caps, one kneeling in a forward sprawling posture, clasping the lug-mast with his arms, his knees wide apart, head thrown back, the yellow eye-balls with their islands of grey iris staring straight up the mast-pole. In every case I found below-decks cruses of corn-brandy, marked "*aquavit,*" two of which I took into the pram; but at one boat, instead of boarding in the pram, I shut off the *Boreal's* liquid air at such a point, that, by delicate steering, she slackened down to a stoppage just a-beam of the smack, upon whose deck I was thus able to jump down; and after looking round I descended the three steps aft into the dark and garrety below-decks, to go with a bent back calling in a sort of whisper: *Anyone? Anyone?;*" but when I went up again the *Boreal* had drifted three yards beyond my reach, so, as there was a dead calm, I had to plunge into the water: and in that half-minute there such a sudden throng of terrors beset me! yes, I can feel again now that abysmal desolation of lonesomeness and sense of a hostile universe bent upon eating me up, for the ocean seemed to me nothing but a great ghost.

Two mornings later I came upon another school, rather larger boats these, which I discovered to be Brittany cod-fishers, and most of these, too, I boarded—in every below-decks a wooden or earthenware image of "the Virgin," painted in gaudy faded daubs, in one boat a boy who had been kneeling before her, but was toppled sideways now, his knees still bent, the cross of Christ in his fist. These blue-woolen blouses and tarpaulin sou'-westers lay in every pose of death, every detail of feature and expression still perfectly preserved; the sloops all the same, all, all: with sing-song creaks they rocked a little, nonchalantly; each, as it were, with a sub-consciousness of its own personality and callous unconsciousness of all the others round it, yet a copy of the others: the same hooks and lines, disemboweling-knives, barrels of salt and pickle, piles and casks of opened cod, kegs of biscuit, and creaky rockings, and a bilgy smell, and dead men. The next day, about eighty miles south of the latitude of Mount Hekla, sighting a big ship, which proved to be the French cruiser *Lazare Tréport,* her, too, I boarded and overhauled during three hours, her upper, main, and armored deck, deck by deck, to her black

Life in the sea -

depths, even peeping up the tubes of her two rusted turret-guns. I saw three men in the engine-room mangled—after death, I presume—by a burst boiler; and I saw about 800 yards to the north-east a long-boat of hers crammed with marines, one oar still there, jammed between the row-lock and the rower's forced-back chin; while on the ship's port deck, in the stretch of space between the two masts, the blue-jackets had been piped up, for there they lay in a sort of serried disorder, two hundred. Nothing could be of a suggestion more tragic than the helpless power of this poor wandering craft, round whose stolid mass myriads of wavelets, busy as aspen-leaves, bickered with a continuous weltering splash which kept chattering loud like sparrow-crowds. I sat a good time that afternoon in one of her casemates on a gun-carriage, my head sunken on my breast, furtively eyeing the blueish turned-up feet, shrunk, bloodless, of a sailor who lay before me, his soles being alone visible, since he lay head downwards beyond the steel door sill; and drenched in seas of lugubrious reverie I brooded there, till, with a shudder, I awoke, got back to the *Boreal*, and till sleep conquered me, went on my way. At nine the next morning, on coming on deck and spying to the west a group of craft, I turned my course upon them, and they turned out to be ten Shetland sixerns, which must have drifted north-eastward hither. I examined them well, but they were as the long catalogue of the others: for all the men, and all the boys, and all the dogs on them were dead.

* * *

I could have come to land a long time before I did; but I would not: I was so afraid. For I was used to the silence of the ice; and I was used to the silence of the sea: but I was afraid of the silence of Europe.

* * *

Once, on the 14th of July, I had spied a whale, or had thought so, spouting remotely afar on the south horizon; and on the 19th I saw a swarm of porpoises vaulting the water in their successive way, northward: and, seeing them, I had said to myself: "Well, I am not alone in the world, then, my good God."

Moreover, some days later, the *Boreal* had found herself in a shoal of cod making away northward, millions of fish, for I saw them, and one afternoon caught three, hand-running, with the hook.

So the sea, at least, had its breeds to be my comrades.

But if the land should be found as still as the sea, without even the spouting whale, or bank of tumbling sea-hogs—if Paris were dumber than the everlasting snow—what then, I asked myself, should I do?

* * *

I could have made short work and landed at Shetland, for I found myself as far westward as longitude 11° 23′ W.; but I would not: I was so afraid. The shrinking within me to envisage that suspicion which I had turned me first to a foreign land.

So I made for Norway, and on the second night of this definite intention, about nine o'clock, the weather being squally, the sky lowering, the air sombrous, and the sea hard-looking, dark, ridged, I was steaming away at a good rate, holding the wheel, my poor port and starboard lights still beaming there, when, without the least notice, I received the roughest shock of my life, being shot bodily as from a cannon to the cabin door, through it head-foremost down the companion-way, and still beyond along the passage, having crashed into some dark ship, probably of large size, though I never saw her, nor any sign of her; and all that night, and the next day till four in the afternoon, the *Boreal* went driving alone over the sea whither she would: for I lay dazed. Then I found that I had received really insignificant injuries, considering; but I sat there an hour on the floor in a sulky, disgusted mood, and, when I got up, pettishly stopped the ship's engines, seeing my dozen dead all huddled and disfigured. Now I was afraid to steam by night, and even in the day-time would not go on for three days: for I was angry with I know not what, and disposed to quarrel with Those whom I could not see.

However, on the fourth day a rough swell which knocked the ship about, and made me uncomfortable, coaxed me into moving, as I did with my bows looking east and south.

I sighted the Norway coast five days later, in lat. 63° 19′, at noon of the 12th of August, and pricked off my course to follow it; but it was with a dawdling reluctance that I crawled, under half-speed. In some eight hours, as I was aware from the chart, I ought to sight the lighthouse-light on Smoelen Island; and when quiet night came, the black lake-water branded with trails of moonlight, I moved close by it, between ten and midnight, almost within the shadow of the mountains: but, Thou God, no shine was there; and all the way down I marked the rugged sea-board slumber darkling, afar or near, with never one friendly light.

* * *

Well, on the 15th of August I had another of those raptures whose passing away would have left an elephant prostrate. During four days I had noted not one sign of present life on the Norway coast, only cliffs, cliffs, dead and dark, and floating craft, all dead and dark; and my eyes now, I found, had acquired a crazy fixity of stare into the abyss of vacancy, while I remained unconscious of being, save of one point, rainbow-blue, far down in the infinite, which passed slowly from left to right before my consciousness a little way, then vanished, he sees a ship under power -

came back, and passed slowly afresh, from left to right continually, until some prick, or voice, would prod me into the consciousness that I was staring, whispering in confidence the warning: "*stare, and all's over with you!*" Well, lost in a trance of this sort, I was leaning over the wheel during the afternoon of the 15th, when it was as if some instinct or premonition sprang up in me to say "If you look yonder, *you will see* . . . !," and in one instant I had ascended from all that depth of reverie to reality, had glanced to the right, and there, at last, my God, I saw something human that moved, at last!—and it came to me.

That sense of recovery, of waking, of new solidity, of the comfortable usual, a millionfold too intense for utterance: anew now I can fancy and feel it—the rocky ordinary, on which to plant the feet, and live: for from the day when I had stood at the Pole, and viewed there the dizzy thing that had made me swoon, there had come into my way not one sign that other things like myself were alive with me, until now, suddenly, I had the proof: for on the south-western sea, not four knots away, I saw a ship, her bows, which were as sharp as a hatchet, briskly chipping through the smooth sea, throwing out profuse ribbons of foam that flowed wide-wavering out, with outward undulations, far behind her length, as she ran the waters in haste, straight northward.

At the moment I was steering about S.E. by S., fourteen knots out from a shadowy-blue series of Norway mountains, and, just giving the wheel one frantic spin to starboard to bring me down upon her, I dashed to the bridge, propped my back upon the mainmast, which passes through it, put a foot on the white iron rail in front of me, and there at once felt all the mocking devils of distracted revelry possess me, as I caught the cap from my hairs, and commenced to wave and wave and wave, maniac that I was: for at a second glance I saw that she was flying an ensign at the main, and a long pennant at the main-top, and I did not know what she was flying those flags there for: and I was embittered and driven mad.

Distinctly did she print herself upon my consciousness in that three minutes' interval, she a dull and cholera yellow, like lots of Russian ships, a space of faded pink visible at her bows under the yellow, her ensign and blue-and-white saltire, a passenger-liner, two-masted, two-funneled, though from her funnels issued no smoke, her steam-cones in all positions, and all about her course the sea spotted with wobbling fulgors of the sun's going down, coarse blots of glory close to the eye, but graduating to a finer pattern in the distance, and at the horizon refined to a line of livid silver.

The double speed of her and of the *Boreal* must have been quite forty knots, and the meeting accomplished within five minutes: yet into that time I crowded years of life, shouting passionately at her insanity, my face and eyes inflamed with rage the most precipitate, uproarious: for she did not slow, nor signal, nor make any show of seeing me, but came furrowing down at me like Juggernaut with a steadfast run, so that I lost reason, thought, memory, sense of relation, in that seizure of hysteria that transported me; and can only remember now that, in the midst of my howling, a sentence howled by the fiends who used my throat to express their frenzy set me laughing high and madly, for I was crying: "Hi! Bravo! Why don't you stop? *Madmen! I have been to the Pole!*"

In that moment an odor arose, and came, and struck upon my brains, most execrable; and while one might count ten I was aware of her engines sounding near, as that cursed charnel went churning the sea past me on her mænad way, hardly twenty yards from my nostrils. She was a thing, my God, from which the vulture would fly with loathing: I got a glimpse of decks piled thick with her festered dead.

Black on her yellow stern my eye-corner caught the word *Yaroslav*, as I bent over the rail to retch and cough and vomit at her: she was a horrid thing.

This ship had without doubt been pretty far south in tropical or sub-tropical latitudes with her throng of corpses: for all the bodies which I had seen until then, so far from smelling ill, seemed to breathe out a certain perfume of the peach; and she was one of those ships which have substituted liquid air for steam, yet retained their old steam-funnels, &c., in case of trouble, for air was still looked at askance by builders on account of the accidents sometimes due to it: so this *Yaroslav* must have been left with her engines working when her crew were overtaken by death, and, her air-tanks being still unemptied, must have been ranging the ocean with impunity ever since, during I knew not how many months, or years.

Well, I coasted Norway for almost a hundred and forty miles without once going closer than two or three miles: for something held me back; but, passing the fjord mouth where I knew that Aadheim was, I suddenly spun the helm to port before I knew that I was doing it, and made for land.

In half an hour I was moving up an opening in the land with mountains on either hand, streaky rock at their summit, umbrageous boscage below; and the whole softened, as it were, by veils woven of the rainbow.

This stretch of water lies as winding as a thread which one drops, only the windings are more pointed, so that every few minutes the scene was a new scene, though the vessel just pushed her way up; and nothing of what was gone behind me could be spied, or merely a land-locked gleam like a pond.

I never saw water so polished, argent, like polished marble, reflecting everything boldly within the womb of its lucid abysm, over which hardly a whiff blew that sundown, wimpling about the bluff *Boreal*, which seemed to move as if shrinking from bruising it, in rich wrinkles and creases, like glycerine, or dewy-trickling lotus-oil: yet it was only the sea; and the grandeur yonder was only crag and autumn-foliage and mountain-slope: yet all seemed caught-up, rapt in a trance of rose and daffodil, compounded of the stuff of dreams and bubbles, of dust-of-flowers, and blushes of the peach.

I saw it not only with joy, but with astonishment, having forgotten, as was natural in all that long barrenness of snow and sea, that aught could be so ethereally beauteous, yet homely, too, human, familiar, and consoling; the air here richly imbued with that peachy odor; and there was a Sabbath and a nepenthe and a charm in that place just then, as it were of those gardens of Hesperus and fields of asphodel reserved for the spirits of the just.

Alas, but I had the glass, and for me nepenthe was mingled with a despair immense as the heavens, my good God: for anon I would take it up to search some perched hut of the peasant, or burg of the "bonder," on the tops: and I saw no one; and to the left, at the fourth bend of the fjord, where there is one of those watch-towers that these people used for watching in-coming fish, I saw on a slope of rock just before the tower a body which looked as if it must tumble headlong; and when I spied him there, I felt definitely, for the first time, that shoreless despair which I alone of men have felt, high as the stars, darkness as hell; and I fell to staring anew that stare of Nirvana and the lunacy of Nothingness, wherein Time merges in Eternity, and all being, like one drop of water, flies dispersed to fill the void of space, and is lost.

The *Boreal's* bow walking over a fishing-boat roused me, and a minute later I saw two people on the shore, which, three feet above the water there, is edged with boulders and shrubs, behind which is a path winding upward through a gorge; and on the path I saw a driver of one of those sulkies called karjolers, he on the high front seat lying sideward and backward, his head resting on the wheel; and on a trunk strapped to a frame on the axle behind was a boy, his head, too, resting sideward on the wheel, near the other's; and the pony pitched forward on its fore-knees, tilting the shafts downward; and a little distance from them on the sea a skiff.

* * *

After the voe's next foreland, I commenced to see craft, whose number increased as I advanced, small boats, with some schooners, sloops, the majority aground; and suddenly now I was conscious that, mingling with that delicious odor of spring-blossoms—profoundly modifying, yet not destroying it—was another odor, wafted to me on the wings of the whiffs from the land; and "Man," I said, "is decomposing:" for I knew it well: the odor of human corruption.

* * *

The fjord opened finally in a wider basin, surrounded by towering mountains that reflected themselves in the basin to their last cloudy crag: at the end of which were ships, a quay, and a homely old town.

Not a sound, not one; only my own engines sluggishly going: and here, it was clear, the Angel of Silence had passed, and his scythe mown.

I ran and stopped the engines, and, without anchoring, got down into a boat that lay at the ship's side, to paddle toward the little quay, passing under a brigantine with her courses set, three jibs, staysails, squaresails, gaff-topsail, looking hanging and listless, and wedded to a copy of herself, mast-downward, with the water; there were three lumber schooners, a forty-ton steam-boat, a tiny barque, five herring-fishers, and ten or twelve shallops: and the sailing-craft had all fore-and-aft sails set; and about each, as I rowed near, brooded an odor which was both sweet and odious, more suggestive of the genius of mortality—the essential mood and meaning of Azrael—than aught that I had ever dreamed: for all, as I saw, were thronged with bodies.

Well, I went up the old mossed steps in that dazed state in which one notices frivolous things: conscious of the lightness of my new clothes, for the day before I had changed to Summer things, having on now only a shirt of undyed wool, the sleeves rolled up, and cord trousers, with a belt, and a cloth cap over my long hair, and an old pair of yellow shoes, without laces, without socks; and from the edge of the quay I looked out over a piece of rough ground which lay between the town and the quay.

What I saw was not only woeful, but wildly startling: woeful, because a multitude had assembled, and lay dead, there; and wildly startling, because something in their *ensemble* informed me in one minute why they were there in such number.

They were there with the motive, and in the hope, to fly westward by boat.

And the something which informed me of this was a certain *foreign* air about that field of dead, as the eye rested on it: something un-northern, southern, Oriental.

Two yards from my foot lay a group of three: one a Norway peasant-girl in a green gown, scarlet stomacher, Scotch bonnet; the second an old Norway man in knee-breeches, "smallclothes," worsted cap; and the third a Jew of the Polish Pale, say, in gaberdine and skull-cap, with ear-locks.

I went nearer to where they lay thick between the quay and a stone fountain in the middle of the space, and I saw among those northern dead two women in costly dress, Spanish or Italian, and the yellower mortality of a Mongol, probably a Magyar, and a big negro in Zouave garb, and some twenty obvious French, and two Morocco fezes, and the green turban of a shereef, and the white of an Ulema.

So I asked myself this question: "How came these foreign stragglers here in this northern townlet?"

And my wild heart answered: "There has been an impassioned stampede, northward and westward, of all the breeds of Man: and this that I here see is but the far-flung spray of that infuriate flood."

* * *

Well, I walked along a street, cautious where I trod, a street not all voiceless, but haunted by swarms of mosquitoes and dreamy twinges and messages of melody at the tympanum, like the drawing of the fiddle-stick in sorrow-land; a street strait, pavered, steep, drear; and the sensations with which I, poor bowed man, went moping about that town, only Atlas, fabled to bear the burden of this earth, may know.

* * *

I thought to myself: If now a swell from the Deep has swept over this planetary ship of earth, and I, who alone chanced to find myself in the furthest stern, as the sole survivor of her crew? . . . What then, my God, shall I do?

* * *

I felt, I felt, that in this place, save the water-gnats of Norway, stirred no living thing; that the hum and the savor of Eternity pervaded, smothered, mummified it.

The houses are mostly of wood, some large, with *porte-cochères* leading into semi-circular yards, round which the buildings stand, steep-roofed in view of the snow-masses of winter; and through one casement of one, near the ground, I saw a stout old woman in a cap on her face before a porcelain stove. But I paced on without stoppage through three streets, and came out, as it got dark, upon a piece of grass-land leading downward to a mountain-gorge, some distance along which gorge it was that I found myself sitting the next morning: and how, and in what trance, I passed all that blank night is obliterated from my mind. When I looked about with the return of light I saw mountains of fir on either side, almost meeting overhead at some points, deeply shading the mossy gorge; and, getting up, careless of direction, I went still onward, to walk and walk for hours, unconscious of hunger, though there was profusion of wild mountain-strawberries, very tiny, which must bloom almost into winter, a few of which I ate; and there were blue gentianellas, and lilies-of-the-valley, and luxuriance of boscage, and always a noise of waters: I saw little cataracts aloft flackering like white wild rags, for they fractured in the mid-fall, and were caught away, and lost; I saw also patches of reaped hay and barley, hung up in a strange way on stakes, I suppose to dry; and perched huts; and a pigmy castle or burg, inaccessible seemingly; but none of these did I enter; and five bodies only I saw in the gorge, a woman with a child, and a man with two small cattle.

Near three in the afternoon, startled to see myself there, I started to go back; and it was dark when I again moved through those gloomy streets of Aadheim, making for the quay, feeling now my hunger and fatigue, without any intention of entering any house; but, as I stepped by one *porte-cochère*, something shoved me in, for my intellect had become as fluff on the winds, not working of its own verve, but the sport of impulses that seemed external: so, after passing across the yard, I ascended a spiral stair of wood by a twilight which just enabled me to pick my way among five or six dim forms fallen there: and in that confined place fantastic qualms beset me. I mounted to the first landing, tried the door, it was locked; mounted to the second: that door was open; and with reluctance, chilly, I took a step inward where all was pitch darkness, the window-stores drawn. I hesitated: it was pretty dark; tried to utter that word of mine, but it came up barely in a whisper, tried still once, and heard myself say: "Anyone?;" but, in venturing yet a step forward, I had trodden upon soft guts, and at that contact terrors got me: for it was as though I beheld the goblin eye-balls of Hell and frenzy goggle upon me out of that gloom; and, murmuring a gurgle of remonstrance, I was gone, helter-skelter down the stairs, treading upon dead, across the yard, down the street, with pelting feet, and open arms, and sobbing bosom, for I thought that all Aadheim was after me; nor was my horrid haste abated till I was on board the *Boreal*, and moving down the fjord.

Out to set, then, I went anew; and within the next few days visited Bergen, and put in at Stavanger: and Bergen and Stavanger were dead.

It was then, on the 20th of August, that I bent my bow toward my native land.

* * *

From Stavanger I steered a straight course for the Humber.

I had no sooner left behind me the Scandinavian coast than I commenced to come among the ships—ship after ship; and by the time I entered the zone of the usual alternation of sunny day and sunless night, I was moving through the medley of an incredible number of craft, a far-cast armada: for over all that expanse of the North Sea, where, in its most populous days of trade, the sailor might perhaps spy a sail or two, I had now at every moment twelve within scope of the glass, oftentimes forty.

And still they lay on a still sea, itself a dead thing, livid as the lips of death, there being a starkness of trance in the calm which was most remarkable: for the ocean seemed weighted, and the air drugged.

Extremely slow was my progress, for at first I would not

leave any ship, however remotely pigmy, without approaching sufficiently to investigate her, at least with the glass; and a multitudinous mixture of species they were, trawlers in hosts, war-ships of every nation, used, it seemed, as passenger-boats, smacks, feluccas, liners, steam-barges, great four-masters with sails, Channel boats, luggers, a Venetian *burchio,* colliers, yachts, *remorqueurs,* training ships, dredgers, two dahabeeahs with curving gaffs, Marsielles fishers, a Maltese *speronare,* American off-shore sail, Mississippi steam-boats, Sorrento lug-schooners, Rhine punts, yawls, old frigates and three-deckers, called to novel use, Stromboli caiques, Yarmouth tubs, xebecs, Rotterdam flat-bottoms, floats, mere gunwaled rafts—anything from anywhere that could bear a human freight on water had come, and was there: and all, I knew, had been making westward, or northward, or both; and all, I knew, were thronged; and all were tombs, listlessly wandering, my God, on the wandering sea with their throngs.

And so fair the world round them: suavest autumn weather; all the atmosphere aromatic with the vernal cherriness of that perfume of peach: yet not so utterly calm, but, if I passed close to the lee of any floating thing, the spicy breathings of morning or evening brought me vague puffs of the odor of the mortal over-ripe for the grave.

So burdensome and accursed did this thing become to me, such a plague and a hissing, vague as was the offense, that I began to shun rather than to seek the ships, and also I now dropped my twelve, whom I had kept to be my companions all the way from the Far North, one by one, into the sea: for now I had definitely passed into a zone of warmth.

I was convinced, however, that the poison, whatever it might be, had some embalming, or antiseptic, effect upon the bodies: at Aadheim, Bergen, and Stavanger, for instance, where the temperature permitted me to go without a jacket, only hints and whiffs of the processes of dissolution had troubled me.

* * *

Very benign, I say, and joyous to see, was sky and sea during all that voyage; but it was at sunset that my sense of the wondrously beautiful was roused and excited, in spite of that burden which I bore: for, certainly, I never saw sunsets resembling those, nor could ever have dreamt of aught so flamboyant, exorbitant and distraught, all the vault seeming transformed to an arena for warring powers warring for the cosmos, or it was like the wild countenance of God routed, and flying flustered through cosmic storm-gulfs from His foes. But many evenings I watched it with unintelligent awe, believing it but a portent of the unsheathed sword of the Almighty, until one morning a thought pricked me like a pin, for I suddenly remembered the wild sunsets of the nineteenth century witnessed in Europe, America, and, I think, everywhere, after the eruption of the volcano of Krakatoa.

And whereas I had previously said to myself "If now a wave from the Deep has washed over this wandering Ship-of-Space . . . ," I said now "A wave—but hardly from the Deep: a wave rather which she had husbanded, and has spouted, from her own unmotherly bowels . . ."

* * *

I had some knowledge of the Morse code, of the manipulation of tape-machines, telegraphic typing-machines wireless transmitting, as of most little things of that sort which came within the outskirts of the interest of a man of science; I had collaborated with Professor Stanistreet in the production of a text-book called *Applications of Science to the Arts,* which had brought us some money: and, on the whole, the *minutiœ* of modern things were still pretty fresh in my memory: so I could have wirelessed, or tried to wire from Bergen, to somewhere; but I would not: I was so afraid; afraid lest for ever from nowhere should occur one replying click, or stir of dial-needle. . . .

* * *

I could have made short work and landed at Hull; but I would not: I was so afraid. For I was used to the silence of the ice; and I was used to the silence of the sea: but I was afraid of the silence of England.

* * *

I came in sight of the coast on the morning of the 26th of August, somewhere about Hornsea, but did not see any town, for I put the helm to port, and went on further south, no longer bothering with the instruments, but coasting at haphazard, now in sight of land, now in the center of a circle of sea, not admitting to myself the motive of this loitering slowness, nor thinking at all, but ignoring the lurking dread of the morrow which I shirked, and furtively dwelling in today; so I passed the Wash, passed Yarmouth, Felixstowe, the things that floated motionless on the sea being now beyond counting, for I could scarcely lower my lids ten minutes and left them without seeing yet another there: so that soon after dusk I, too, had to stand still among them all, until morning, for they lay dark, and to cruise about would have been to drown the already dead.

Well, I came to the Thames-mouth, and lay pretty well in among the Flats and Pan Sands toward nine one evening, not seven miles from Sheppey and the North Kent coast: and I did not see any Nore Light, nor Girdler Light; and all along the coast I had seen no light, though as to that I breathed not a syllable to myself, not admitting it, nor letting my heart know what my mind argued, nor my mind know what my heart surmised; but with a mock-mistrustful underlook, half daft, I would regard the darkling land, considering it a sentient thing that would be playing a prank upon a poor man like me.

And the next morning, when I idled further on, my furtive eye-corners were very well aware of the Prince's Channel light-ship, also of the Tongue ship, for there they were; but I would not look at them at all, nor steer near them, for I did not want to have anything to do with whatever might have happened beyond my own ken, and it was better to look straight before, observing nothing, and concerning one's-self with one's-self.

The next evening, after having gone out to sea again, I was again in, a little to the E. by S. of the North Foreland, and I saw no light there, nor any Sandhead light, but over the sea vast signs of wreckage, and the coasts were strewn with old wrecked fleets; then I moved away about S.E., very slowly steaming—for anywhere hereabouts hundreds upon hundreds of hulls lay dead within a ten-mile circumference of sea—and by two in the 'foreday had roamed up well within sight of the French cliffs: for I had said "I will go and see the light-beam of that revolving-drum on Calais pier, that nightly beams half-way over-sea to England;" and the moon shone clear in the southern sky that morning, like an old dying queen whose Court swarms distantly from round her diffident, pale, tremulous, the paler the nearer; and I watched the mountain-shadows about her spotty full-face, and her nimbus of mist, and her beams on the sea, as it were kisses sneaked in the kingdom of sleep, and among the quiet ships white trails and powderings of light, strange, agitated, like palace-corridors in some fairyland forlorn, thronged with wan whispers, scandals, and runnings-to-and-fro, with leers, and breathless last embraces, and flight of the princess, and death-bed of the king; and on the N.E. horizon a streak of cloud that seemed outside the sky; and yonder, not far, the chalk coast-cliffs, not so low as at Calais near, but arranged in masses with vales of sward between, each with its wreck: but no beam of any revolving-drum I saw.

* * *

I could not sleep that night: for all the operations of my mind and body seemed in abeyance: so, mechanically, I moved the ship westward once more, until the sun came up, when, scarcely two miles from me, there stood the cliffs of Dover, and over the crenulated summit of the Castle I noted the Union Jack hanging motionless.

I heard eight, nine, o'clock strike in the cabin, and I was still at sea; but some audacious whisper was at my brain: and at 10.30, the 2nd of September, just opposite the Cross Wall Custom House, the *Boreal's* anchor-chain, after a voyage of three years, two months, and fourteen days, ran thundering, thundering, through the starboard hawsehole.

Ah, Heaven! but I must have been made to let the anchor go! for the effect upon me for that obstreperous hubbub, breaking out sudden upon all that cemetery repose that blessed morning, and bellowing, it seemed, a year, was appalling; and at the cruel racket I stood excruciated, shivering with a flinching heart, God knows: for not less uproarious than the rumpus of the Judgment trump it raged and raged, and I thought that all the armies of the dead could hardly fail to start and rise at alarum so excessive, and question me with their eyes. . . .

* * *

On the top of the Cross Wall I saw a crab crawling; at its end, where a street begins, I saw a gas-light, and at its foot a black man on his face, clad in a shirt and one boot; I saw the harbor packed with all sorts of craft, and on a Calais-Dover boat nine yards from me I saw the dead piled, she being unmoored, and continually grinding against a green brig.

And when I saw that, I dropped down there by the capstan, and my heart sobbed, as I said "Well, Lord God, Thou hast destroyed the work of Thy hand. . . ."

* * *

After a time I got up, went below in a state of somnambulism, took a packet of pemmican cakes, leapt to land, and went following the railway that runs from the Admiralty Pier to a passage with railway-masonry on one side, in which I saw five dead, and could not believe that I was in England, for all were dark-skinned people, three gaudily dressed, two in flowing robes; and the same when I walked into a street leading northward, for here were a hundred, and never saw I, except in Constantinople, where I once lived eighteen months, so variegated a mixture of races, black, brunette, brown, yellow, white, some emaciated like people dead from famine; and, over-looking them all, one boy in an Eton collar seated on a bicycle, supported by a lamp-post which his arms clasped, he proving the extraordinary suddenness of the death which had overtaken them all.

I did not know whither, nor why, I went, nor had I any notion whether all this was palpably beheld by me in the planet which I had known, or in some other, or was all phantasy of my disembodied anima, for I had the thought that I also might be dead since old ages, my soul roaming now through the profoundness of space, in which there is neither north nor south, nor up nor down, nor measure nor relation, nor aught whatever, save an uneasy consciousness of a dream about bottomlessness. Of sorrow or pain, I think, I felt nothing, though I have a sort of memory now that some sound resembling a sob or groan, though it was neither, proceeded at regular intervals from my bosom during three or four days. Meantime, my brain registered like a tap-machine details the most frivolous—the name of a street, Strond Street, Snargate Street; the fur cap—black fur for the side, ermine for the top—of a portly Karaite priest on his back, his robes blown up to his knees, and neatly folded there; a violin-bow gripped between the irregular teeth of a little Spaniard, his hair brushed back, mad-looking eyes; odd shoes on the feet of a French girl, one black, one brown: they lying as numerous as gunners who fall round their carriage, five to ten feet apart, the majority, as also in Norway and on the crafts, in postures of distraction, with far-spread arms, frantic distortion of limb, like men who in the instant before death called upon the rocks and hills to cover them.

* * *

I came to an opening in the land, named, I think, "The Shaft," into which I passed, climbing a great number of steps, which I began to count, but left off, then the dead, and left off; and finally, at the summit, which must be even higher than the Castle, came to a great space laid out with gravel-walks, and saw fortifications, barracks, a citadel. I was surprised at the breadth of view. Between me and the Castle to the east lay the crowd of houses, brick and rag-stone, mixed in the distance with a vagueness of azure haze; and to the right the harbor, the sea, the ships; and about me on the heights nine or ten dead, biting the dust; the sun now high, warm, with hardly a cloud in all the vastness of the vault; and yonder a cloud that was the Norman coast. It seemed too big for one poor man.

My head nodded, I sitting on a bench of boards, with intervals; and, as I saw it all, I nodded, heavy-headed, weary: for it was too big for me; and, as I nodded, my forehead propped on my left hand, there was in my head an old street-song that I groaned sleepily, like coronachs and dread funereal nenias, the packet of pemmican-cakes beating time in my right hand, rising and dropping, dropping heavily and rising, in time . . .

I'll buy the ring,
You'll rear the kids:
There'll be servants to wait on our ting, ting, ting.

. . . .

Ting, ting,
Won't we be happy?
Ting, ting,
That shall be it;
I'll buy the ring,

You'll rear the kids:
There'll be servants to wait on our ting, ting, ting.

. . . .

So, maundering, I dropped forward upon my face; and for twenty-three hours, the living undistinguished from the dead, I slept there.

* * *

I was awaked by drizzle, leapt up, and, on looking at my silver chronometer, which, attached by a leather to my belt, I carried in my trousers' pocket, saw that it was 9 a.m., the sky now sombrous; and a moaning wind—almost a new thing now to me—had risen.

I ate some pemmican, for I had a reluctance—needless, as it proved—to eat any of the thousand luxuries here, sufficient no doubt, in a town like Dover alone, to last me hundreds of years; and, having eaten, I descended "The Shaft", to spend the whole day, though it rained and blustered, in strolling about. Reasoning in my numb way from the number of ships on the sea, I believed that the town would be found to be over-crowded with dead, but this was not so, for that westward furore and stampede must have operated here also, leaving the town empty but for the new-coming hosts.

My first work was to go into a grocer's shop, which was a post-and-telegraph office, with the notion, I suppose, to get a message through to somewhere, in this shop a single gas-jet glimmering its last, this and that other near the pier being the only two which I saw: and garishly enough they glared there, transparently wannish, as it were shamed, like blinking night-things surprised by the brilliance of day, they having so flared and stared for months, or years, inasmuch as they were now bazing diminished, with streaks and rays in the flame, as if by effort: so, if these were the only two, months must have been needed almost to exhaust the gasometer; and this gas-jet blinked upon a negro with a number of parcels scattered about him, and on the counter an empty till, and behind it a little woman, her face resting sideways in the till, her fingers clutching the outer counter-rim, with such an expression of terror! So I got over the counter to a table behind a wire-gauze, and went over the Morse alphabet in my mind before touching the Wheatstone drop-handle, never asking myself who was to answer my message, habit being still strong upon me, and my mind dodged from reasoning from what I saw to what I did not see; but when I moved the commutator, and peered at the dial-needle at my right, as it did not move, I knew that no current was passing, for one pair of the commutator-spikes had apparently been in contact with a pair of the uprights, so the battery had run down: and with a kind of fright, I was up, leapt, and got away from the place, though there was a number of telegrams about, which, if I had been in my senses, I would have read.

At the next street-corner I saw open the door of a large house, and went in: but from bottom to top no one there, except one English girl, seated in an easy-chair in a drawing-room furnished with Valenciennes curtains and azure-satin, a girl of the "submerged" class clad in rags, and there she lay back with a hanging jaw in an awkward sort of posture, a jemmy at her feet, she clutching a lot of bank-notes, in her lap two watches: in fact the bodies here were either those of foreigners, or else of the very poor, the very old, or the very young.

But what made me remember that house was that I found there on a sofa a paper, *The Kent Express;* and, sitting unconscious of my neighbor, I pored long over what was written there.

It said in an article that I tore out and kept: "Communication with Tilsit, Insterburg, Warsaw, Cracow, Przemysl, Gross Wardein, Karlsburg, and many smaller towns immediately east of the 21st of longitude has ceased during the night, though in some at least of them there must have been operators still at their posts, undrawn into the westward-rolling torrent: but as all messages from Western Europe have been met only by that mysterious muteness which, three months and two days since, astounded civilization in the case of Eastern New Zealand, we can only assume that these towns, too, have been added to the mournful catalogue; indeed, after last evening's Paris news we might have foretold with some assurance, not merely their overthrow, but even the moment of it: for the rate of the slow-riding vapor which is touring our globe is no longer doubtful, having now been definitely fixed by Professor Craven at 100½ miles a day—4 miles 330 yards an hour. Its nature, its origin, remain matters of conjecture: for it seems to leave no living thing behind it; nor, God knows, is that of any moment now to us who remain. The rumor that it is associated with an odor of almonds is asserted on good authority to be improbable, but the morose purple of its impending gloom has been attested by tardy fugitives from the face of its rolling and smoky march.

"Is this the end? We do not, will not, believe it. Will the sweet sky which today smiles over us be invaded in nine days, or less, by this smoke of Night? In spite of the avowals of the scientists, we still doubt. For, if so, to what purpose that drama of Evolution in which we seem to see the artistry of the Dramaturgist? Surely, the end of a fifth act should be obvious, satisfying to one's sense of the complete: but History so far, hoary as it has been, resembles rather a prologue than a fifth act. Can it be that the Manager, utterly dissatisfied, would sweep all off, and 'hang up' the piece for ever? Certainly, the sin of mankind has been as scarlet: and if this Heavenly earth that he has converted into Hell smother him now under the muck of Hell, little the wonder. But we will not yet believe. There is a sparing strain in Nature; through the world, as a thread, is spun a silence which smiles; and on the end of events we find placarded large the words: 'Why were ye afraid?' A tranquil hope, then—even now when we crouch beneath this world-wide shadow of the wings of the bird of death—befits us: and, indeed, we see such an attitude among some of the humblest of our people, from whose heart arises the sigh, 'Though He slay me, yet will I trust in Him.' Hear, therefore, O Lord; O Lord, look down, and save.

"But even as we thus speak of hope, reason, if we would hear her, whispers us 'dreamer': and inclement is the sky of earth. No more craft can New York harbor hold, and whereas, among us, men perish of privations by the hundred thousand, yonder across the sea they perish by the million: for where the rich are pinched, how can the indigent live? Already 850 out of the 1500 millions of our race have perished; and the empires of civilization have crumbled like sand-castles to an encumbrance of anarchies. Thousands of unburied dead, anticipating the more deliberate doom that comes and smokes, and rides and comes and comes, and does not tire, strew the streets of London, Manchester; the guides of the nation have fled; the husband stabs his wife for a slice of bread; the fields lie waste; crowds carouse in our churches, universities, palaces, banks, hospitals; we understand that late last night three territorial regiments, the Munster Fusiliers, and the Lothian and East Lancashire Regiments, riotously disbanded themselves, shooting two officers; disease, as we know, is come into its kingdom; in several towns the police seem to have disappeared, and, in nearly all, every vestige of decency; the results following upon the release of the convicts appear to be monstrous in the respective districts; and within three months Hell seems to have acquired this planet, sending forth Horror, like a wolf, and Despair, like a disastrous sky, to devour and confound her. Hear, therefore, O Lord, and forgive our iniquity; O Lord, we beseech Thee; look down, O Lord, and spare."

* * *

When I had read this, and the rest of the paper, which had one sheet-side blank, I sat an hour there, eyeing a patch of the purple ash on the floor close to where the girl sat with her timepieces in her eternity; and there was not a feeling in me, except a pricking of curiosity, which later became morbid, to know more about that cloud of smoke of which this paper spoke, of its dates, its source, its nature; then I went down, and entered several houses, seeking for more papers, but did not see any; then found a paper-shop which was open, with notice-boards outside, but either it had been abandoned, or

Running a train out from Dover.

printing must have stopped near the date of the paper that I had read, for the three papers there were dated long previously, and I did not read them.

Now it was raining, and a blustering autumn day it was, distributing all the odors, continually bringing me mixed whiffs of blossoms and the stench of decay; but I would not mind it much, wandered and wandered, till I was tired of spahi and bashi-bazouk, of Greek and Catalan, of Russian "pope" and Coptic abuna, of dragoman and Calmuck, of Egyptian maulawi and Afghan mullah, Neapolitan and sheik, and the nightmare of wild poses, colors, stuffs and garbs, yellow-green kefies of the Bedouin, shawl-turbans of Baghdad, the red tarboosh, the voluminous rose-silk tob of women, and face-veils, the laborer's corduroy, and stark distorted nakedness, and sashes of figured muslin. About four I found myself seated for very weariness on a doorstep, bent beneath the rain, but soon was up anew, fascinated may-be by this changing bazaar of sameness, its chance combinations and permutations, its novelty in monotony, and about five was at a station marked Harbor Station, in and about which lay a crowd, but no train. There I sat again and rested, rose and roamed again, until after six I found myself at another station named "Priory"; and here I saw two long trains, both bethronged, one on a siding, and one at the up-platform.

On examining both engines, I found them of the old steam-type, in one no water, but in that at the platform the gauge showed some; and, on overhauling all the machinery, I found it good, though rusted, with plenty of fuel, of oil, which I supplemented from a shop near; and for ninety minutes my mind and hands acted with an intelligence as it were automatic, till I saw the fire blazing finely, the steam-gauge registering; and when the safety-valve lever, whose load I lightened by two atmospheres, lifted, I jumped down to try to disconnect the string of carriages from the engine, but failed in this, the coupling being some automatism new to me; nor did I care. As it was now dark, and still some oil for bull's-eye and lantern, I lit them; then rolled driver and stoker, one to the platform, one upon the rails; and about 8.30 ran out from Dover, my throttle-valve pealing high a long falsetto through the bleak and desolate night.

* * *

My aim was London; but I knew nothing of the metals, their junctions, facing-points, sidings, shuntings, and complexities, nor was even sure whether I was raging toward, or away from, London; but just in proportion as my timorousness of the engine hardened into familiarity and self-confidence, I quickened speed, wilfully, with an obstinacy deaf and obdurate, till finally, from a crawl, I was flying at a shocking velocity, while something, tongue in cheek, seemed to whisper me "there must be trains blocking the rails, at stations, in sheds, everywhere—it is a maniac's ride, a ride of death, Flying Dutchman's frenzy; remember your dark brigade of passengers who rock and bump together, and will suffer in a shock"; but stubbornly I thought "they wished to go to London": and on I raged, not crazily exhilarated, I think, but feeling a wicked and morose unreason glow dully in my bosom, while I stoked begrimed at the fire-box, or caught sight of the corpse of horse or ox, of trees and fields receding, glooming homestead and farm, flowing ghostly past me.

Long, though, it did not last: I could not have been twenty miles from Dover when, on a straight stretch of line, I made out before me a tarpaulined mass opposite a signal-box: and instantly callousness popped into panic in me. But even as I put on the brake, dragged at the link-gear lever, I understood that it was too late—rushed toward the gangway for a wild jump down an embankment to the right, but was flung forward by a series of rough bumps, caused by some ten oxen that lay there across the rails; and when I picked myself up and leapt, some seconds before the collision, the speed must have slackened, for I received no fracture, but lay in semi-coma in a patch of yellow-flowering whin on level ground, just conscious of a conflagration on the rails forty yards away, and, all the dark hours, of vague thunder sounding from somewhere.

* * *

By five in the morning I was sitting up, rubbing my eyes, seeing in a dim light mixed with drizzle that the train of my last night's debauch was a huddled-up chaos of carriages and bodies, while on my right a five-barred gate swung with groans; and four yards from me a wee pony with a swollen wan belly, the picture of death; and dead wet birds.

I picked myself up, to go through the gate up a row of elms to a house which I found to be a tavern with a barn, forming one house, the barn much larger than the tavern part; and I went into the tavern by a side-door—behind the bar—into a parlor—up a little stair—into two rooms, but no one there; then round into the barn, paved with cobblestones, and there lay a mare and foal, some fowls, two cows; then up a ladder-stair to a trap-door, and on the floor above in the middle of a wilderness of hay saw nine laborers, five men and four women, huddled together, with some spirit in a tin-pail, so that these had died riotous.

Amid them I slept three hours, afterwards went back to the tavern, and had some biscuits, of which I opened a new tin, with some ham, jam and apples, of which I made a good meal, for my pemmican was gone.

Afterwards I went following the rail-track on foot, the engines of both the trains in collision being smashed, knowing north from south by the sun; and, after many stoppages at houses, arrived, about eleven in the night, at a populous town.

By the Dane John and the Cathedral I recognized it as Canterbury, which I knew well, and walked up to the High Street, conscious for the first time of that regularly-repeated sound, like a sob or groan, which was proceeding from my throat. As there was no visible moon, and these old streets pretty dim, I had to pick my way, lest I should desecrate the dead with my foot, and they all should rise with hue-and-cry to hunt me. However, the bodies here were not numerous, most, as before, being foreigners: and these, scattered about this prim old city in that mourning darkness, presented such a spectacle of the baleful wrath of God, as broke me quite down at one place, where I stood constrained to jeremiads and sore sobbings, crying out upon it all, God knows.

"Not numerous"—till I stood at the west entrance of the Cathedral, whence I could descry spreading up the darkling nave, to the lantern, to the choir, a phantasmagorical mass of forms; and, going a little inward, flashing three matches, peering nearer, I seemed to see the transepts, too, crowded, the south-west porch thronged, so that a great congregation must have flocked hither shortly before their doom overtook them.

Here it was that I became convinced that the after-odor of the poison was not simply lingering in the air, but was being more or less given off by the bodies: for the blossomy odor of this church positively submerged that other odor, the whole rather breathing the aroma of old moldy linens embalmed for years in cedars.

Well, with a stealthy trot I was off from the abysmal stillness of that place, but in Palace Street near made one of those immoderate rackets which seemed to outrage the creation and left me faint, breathless—the racket of the train being different, for there I was fleeing, but here a captive, and which way I fled was capture: for, passing along Palace Street, I saw a lamp-shop, and, wanting a lantern, attempted to get in; but the door was fastened, so, after going away, and kicking against a policeman's staff, I went again to fracture the window-glass—knew that it would make a row, and for ten minutes stood hesitating; but never could I have expected *such* a row, so passionate, dominant, divulgent, and, O Heaven, so long-lasting: for I seemed to have struck upon the weak spot of some planet which came tumbling, with protracted racket and débâcle, about my brows. It was an hour before I would climb in, but then found what I wanted, and some oil-cans; and until two in the morning the innovating flicker of my lantern went peering at random into the nooks of the town.

Under an arch that spanned an alley I saw the window of a little house of rubble, and between its sashes rags beaten in to make the place air-tight against the poison; but when I went in I found the door of that room open, though it, too, had been stuffed at the edges, and on the threshold an old man and woman lay low: so I conjectured that, thus protected, they had remained shut in, till hunger, or the lack of oxygen, drove them forth, whereupon the poison, still active, must have ended them; and I was to see later that this expedient of making air-tight had been widely resorted to, though the supply both of inclosed air and food had nowhere proved commensurate with the duration of the poisonous state.

Weary as I became, some morbid persistence sustained me, and I would not rest, so that four in the morning found me at a station afresh, industriously stooping, poor wretch, at the sooty task of getting another engine ready for travel: for nowhere hereabouts did I see any motor-cars, all having fled westward; and this time when steam was up I succeeded in uncoupling the carriages from the engine: so by the time daylight glimmered I was gliding light away over the country, whither I did not know, but thinking of London.

* * *

Now I went with more wariness, and got on very well, travelling seven days, seldom at night, never at more than twenty miles, slowing in tunnels. I do not know into what maze the train took me, for soon after leaving Canterbury it must have shunted down some branch-line, nor did the names of stations help, for their situation relatively to London I seldom knew; and again and again was my progress interrupted by trains on my metals, when I would have to run back to some shunting or siding; in two instances, these being remotely behind, I transhipped from my own to the impeding engine. On the first day I travelled unimpeded till noon, when I drew up in open country that seemed uninhabitated for ages, only that half a mile off on a shaded sward was a house of artistic design, coated with tinted harling, the roof of red Ruabon tiles, with timbered gables, and I walked to it after another to-do with putting out the fire and laying a new one, the day lightsome and mild, with counties of white cloud lying quiet over the sky. I found in the house an outer and an inner hall, oil-paintings, a kind of museum, in a bedroom three women with servants'-caps and a footman arranged in a strange symmetrical way, head to head, like rays; and, as I stood looking at them, I could have sworn, my good God, that someone was coming up the stair—some creaking of the breeze in the house, increased a hundred-fold to my fevered hearing: for, used to this muteness of eternity that I have heard for years now, it is as though I hear sounds through an ear-trumpet. So I went down quick, and, after eating, and drinking some clary-water, made of brandy, sugar, and rose-water, which I found in plenty, I lay down on a sofa in the outer hall, and slept until midnight.

I went out then, still possessed with the greed to reach London: and, after getting the engine to rights, went off beneath sparkling black sky swarming with spawn of stars far-cast, some of them, I thought, not unlike this of mine, whelmed in immensity of silence, with one life perhaps to see it, and hear its silence; and all the night I travelled, stopping twice only, once to get the coal from an engine which had blocked me, and once to drink some water, which I took care, as always, should be running water. When I felt my head nod about 4 a.m., I tossed myself, just outside the arch of a tunnel, upon a bank thick with stalks and flowers, the workings of early dawn being then in the east: and there, till near eleven, slept.

On waking, I noticed that the country now looked more like Surrey than Kent—that regular swelling of the land; but in fact, though it must have been either, it looked like neither, for already everything had an aspect of tending to a state of wild nature, and I could well divine that for a year at the least no hand had tended the land—close before me being a few roods of lucerne of such superlative luxuriance, that I was led during that day and the next to scrutinize the state of vegetation with some minuteness, and everywhere detected a certain tendency to hypertrophy in stamens, calycles, pericarps, pistils, in every sort of bulbiferous thing that I looked at, in the rushes, above all, the fronds, mosses, lichens, and all cryptogamia, and in the trefoils, clover especially, and some creepers. Many crop-fields, it was clear, had been prepared, but not sown, some not reaped, and in both cases I was struck with their aspect of rankness, as also in Norway, and was all the more astonished that this should happen in the months when a poison whose action is the arrest of oxidation had traversed the earth; I could only conclude that its presence in voluminous masses in the lower strata of the atmosphere had been more or less temporary, and that this tendency to exuberance that I noticed must be due to some principle by which Nature acts with freer energy and larger scope in the absence of man.

Two yards from the rails I saw when I stood up a rill at the foot of a rotten bit of fence, barely oozing itself onward under masses of stagnant fungoids; and here there was a sudden splash, and life, I catching sight of the hind legs diving of a frog; so, going to lie on my belly to pore over the wobbling little water, I presently saw three bleaks or ablets go gliding tiny, low down among the moss-hair flying wild from the bottom-rocks, and I thought how gladly would I become one of them, with my home so thatched and shadowy, and my life drenched in their wide-eyed reverie. At all events, these little beings are living, the batrachians also, and, as I found the next day, chrysalides of one sort or another, for, to my profound emotion, I saw a little butterfly staggering in the air over the flower-garden of a rustic station named Butley.

* * *

It was while I was lying there, poring upon that brooklet, that a thought arose in me: for I said: "If now I be here alone, alone, alone . . . alone, alone . . . one on the earth . . . and my girth have a span of 25,000 miles . . . what will happen to my mind? Into what kind of creature shall I writhe and change? I may live two years so! What will have happened then? I may live five years—ten! What will have happened after the five? the ten? I may live twenty, fifty . . ."

Already, already, there are things that peep and spring within me . . . !

* * *

Wanting food and fresh running water, I walked from the engine through fields of clover whose luxuriance concealed the footpaths, and reached my shoulders; and, after turning the shoulder of a hill, came to a park, in passing through which I saw some deer and three persons, then emerged upon a lawn with terraces, beyond which stood an Early English house—brick with copings and stringcourses of limestone, and spandrels of carved marble: before the porch being a table, or series of tables, in the open air, still spread with cloths that resembled cerements after months of burial, the table having foods on it and some lamps, and all round it, and on the lawn, rustics. I seemed to know the house, no doubt from some print, but could not make out the escutcheon, though I could see from its simplicity that it must be ancient; and over it across the façade spread still some of the letters in evergreens of "Many happy returns of the day": so that someone must have "come of age," or something, for here all was joyance, and it was clear that these people had defied a doom which they foreknew. I went almost through the spacious place of halls, marbles, famous oils, antlers, arras, placid bed-chambers, and it took me an hour. In one of a vista of three reception-rooms lay what must have been a number of quadrille-sets, for to the *coup d'œil* they presented a two-and-two look, made very repulsive by their jewels: and I had to steel my heart to go through this house, for I did not know if these people were looking at me as soon as my back was turned. Once I was on the point of flying, for, as I was stepping up the central stairway, there came a pelt of dead leaves against a window-pane in the corridor above, which thrilled me to my soul; but I thought that, if I once fled, they would all be at me from the rear, and I should be gibbering shrill ere I reached the outer hall, so stood my ground, even challengingly advancing; and in a small dim bedroom in the north wing saw a tall lady, with a groom, or woodman, riveted in an embrace on a settee, she with a coronet on her forehead, their lipless teeth still greedily pressed together. Then I collected in a bag some delicacies from the under-regions, salami, mortadel, apples, roes, raisins, biscuits, some wines, bottled fruit, coffee, and so on, with tin-opener, fork, etc., and dragged them all the way back to the engine before I could eat.

* * *

My brain was in such a way, that it was days ere the obvious means of making my way to London, since I wanted to go there, got into my head, so that the engine went wandering the intricate railway-system of the south-country, I having twice to water her with a coal-bucket from ponds: for the injector was giving no water from the tank, and I did not know where to look for tank-sheds. On the fifth evening, instead of into London, I ran into Guildford.

* * *

That night, from eleven till the next day, a great gale reigned over England: let me note it down; and ten days later, on the 17th, came another; on the 24th another; and I should find it hard now to count the number since: and they hardly resemble British storms, but rather Arctic storms in a certain remarkable something of personalness, and a carousing rowdiness, and a Tartarus dark, that I can hardly half describe. That night at Guildford, after butting about and getting very tired, I threw myself upon a pew in a Norman church with two east apses, called St. Mary's, using the pulpit-cushion for pillow; a little lamp, turned low, burned some distance from me, whose ray served me for *veilleuse* through the night, only one old dame in a chapel on the south side of the chancel, whom I mistrusted, being there with me; and there I lay hearkening, for after all I hardly slept, while over me vogued the megaphones of the immense tempest. Happily I had taken care to close up everything, or, I feel sure, the roof must have gone; and I communed with myself, thinking: "I, poor man, lost in this conflux of infinitudes and vortex of Being, what can become of me, my God? For dark, ah, dark, is this void into which from solid ground I am now gone a trillion furlongs down, the toy of all the whirlwinds: and it would have been better for me to have deceased with the dead, and never to have seen the tenebrousness and turbulence of the ineffable, nor to have heard the thrilling bleakness of the winds of eternity, when they yearn, and plead, and whimper, and when they vociferate and blaspheme, and when they reason and intrigue and entreat, and when they despair and faint, which ear should never hear: for they mean to eat me up, I know, these vast darks, and soon like chaff I shall pass, leaving this scene to them;" so till the morning I lay mumping, with shudderings and cowerings: for the shocks of the storm pervaded the locked nave to my heart; and there were hubbubs of thunder that night, my God, like callings and laughs and banterings bawled across from hill-top to hill-top in Hell.

* * *

Well, in the morning, going down the steep of the High Street, I found a young nun at the bottom whom I had observed the previous evening with a troop of girls in uniform opposite the Guildhall half-way up the street, she having been spun down arm-over-arm; and whereas I had left her dressed to her wimple and beads, she was now stripped, and her little crowd slung about; and boughs of trees fleeing, and huddled houses, and clouds of leaves reeling, were all about me that bleak morning.

This Guildford begin a junction, before again setting out in the afternoon when the gale had lulled, having got an A.B.C. and a railway-map, I decided upon my line, feeling certain now of making London, only thirty miles away; and about five o'clock was beyond Surbiton, expecting every minute to see the city, until night fell; and still, at considerable risk, I went, as I thought, forward: but no London was there, I having, in fact, been on some loop-line, and beyond Surbiton gone wrong again: for the next nightfall I found myself at Wokingham more remotely away than ever.

There I slept on a mat in the passage of a tavern named The Rose, for there was a mad Russian-looking man with projecting teeth on a bed in the house, whose appearance I did not like, and I too tired to walk further; and, setting out early again the next morning, at 10 a.m. I was at Reading.

The notion of navigating the land by the same means as the sea, natural as it was, had not occurred to me; but at the first sight of a compass in a shop-window near the river at Reading, my difficulties as to reaching any particular place vanished: for a chart or map, the compass, a divider, and, in the case of long distances, a quadrant, were all that were necessary to change an engine into a land-ship, one choosing the lines which ran closest to one's course whenever they did not run precisely.

Thus provided, I ran out from Reading in the evening, while there was still some light, having spent there nine hours, this being the town where I first observed that crush of humanity which I afterwards met in towns west of London, the English here quite equal in number to the foreigners, and enough of both, God knows: houses in every room of which, and on the stairs, the dead overlay each other, and in the streets points where only on bodies was it possible to step. I went into the County Gaol, from which, as I had read, the prisoners had been set free, and there found the same crowdedness, cells occupied by ten, corridors rough-paved with faces and old-clothes-shops of robes; and in the parade-ground, against one wall, a mass of stuff, like tough grey clay mixed with rags and trickles of gore, where a cram as of hydraulic horse-power must have acted. At a corner near the buscuit-factory I saw a boy, whom I believe to have been blind, standing jammed, on his wrist a chain, at the end of the chain a dog, he in a haphazard posture from which I conjectured that he and chain and dog had been lifted and placed so by the storm of the 7th; and what made it odd was that his arm pointed rather outward over the dog, so that he looked a drunken fellow setting his dog at me; indeed, all the dead were very mauled and flurried by the storm, and the earth seemed to be making an abortive effort to sweep her streets.

Well, some way out from Reading I found a flower-seed farm looking dead in some plots, in others flourishing rank; and here afresh, fluttering near the engine, three little winged aurelians in the evening air. After which I passed crowds of crowded trains on the down-line, two in collision, even the fields on either hand having a populous look, as if people, when trains and vehicles failed, had set to tramping westward in caravans and streams.

On coming to a tunnel close to Slough, I remarked round the foot of the arch a mass of wooden *débris,* and, as I moved through, was alarmed by the bumping of the engine jumping across bodies; at the other end more *débris;* and I supposed that a company of desperate folk had made the tunnel air-tight at the two arches and provisioned themselves, in the hope to live there till the day of destiny was ended; whereupon their barricades must have been crashed through by some up-train and themselves crushed; or else, other crowds, crazy to share their cave of refuge, had battered down the boardings: this latter, as I afterwards found, being an everyday event.

I should soon have come to London now, but, by bad luck met an up-train on the metals with not a soul in it, and there was nothing to do but to tranship with all my things to its engine, which I found in good condition, with coal and water; and I set it going—a hateful back-ache, I already black from hair to toes. However, by half-past ten, when I found myself stopped by another train, I was only four hundred yards from Paddington, and walked the rest of the way among trains within which the dead still stood upright, propped by one another, and over rails where bodies were as ordinary and cheap as waves on the sea, or twigs in a forest: for throngs had given chase on foot to moving trains, or forerun them in the frenzied hope of inducing them to stop.

I came to the great shed of glass and girders which is the station, the night perfectly soundless, moonless, starless, the hour about eleven; and now I saw that trains, in order to move at all, must have moved through a slough of bodies that had been pushed from behind, and formed a packed mass on the metals; and I knew that they *had* moved; nor could *I* now move, unless I decided to wade, for flesh was everywhere, on the roofs of trains, cramming the intervals betwixt them, on the platforms, splashing the pillars like spray, piled upon lorries, a carnal marsh; outside, too, it filled the intervals betwixt an army-park of vehicles, carpeting that district of London; and all here that odor of blossoms, which nowhere yet, save on one sickening ship, had failed, was now overcome by another.

I found later that all the generating-stations that I visited must have been shut down prior to the arrival of the doom, also that gasworks had been abandoned some time prior: so that this city of dreadful night, in which, at the moment when silence choked it, not less than twenty millions swarmed and droned, must have more resembled the shades of Orcus than aught to which my imagination can compare it.

I got out from the station, with ears, God knows, that still awaited the accustomed noising, but, habituated as I now was to that void of soundlessness, I was overwhelmed in a new awe, when, instead of lights and wheels rolling, I saw the long street which I knew brood lugubrious as Babylons grass-grown, and heard a shocking silence, uniting with the silence of those lights of eternity on high.

* * *

I could not drive any vehicle for some time, for all thereabouts was practically a block; but near the Park, which I attained by stooping among wheels and selecting my foul steps, I boarded a brougham, found in it petrol, set the lamps burning, removed with averted abhorrence four bodies, mounted, broke that populous dumbness, and through streets nowhere empty of bodies went humming eastward my bumpy and besputtered way.

* * *

That I should have persevered, with so much trouble, in coming to this unbounded catacomb, now seems fantastic of me: for by that time I could hardly have expected to find any other like myself, though I cherished, I remember, the (irrational) hope of yet somewhere finding dog or cat, and would anon think bitterly of Reinhardt, my Arctic dog, which my own hand had shot; but, in reality, a curiosity must have been in me to read the real facts of what had happened, so far as it was known or guessed, and to gloat upon all that drama, and cup of trembling, and pouring out of the vials of wrath, in the months prior to the arrival of the end of time—a curiosity which had everywhere made the hunt for papers uppermost in my thoughts; but I had found only four, all antedated to the one that I had read at Dover, though their dates gave me some idea of the period when printing must have ceased, about the 17th of July, three months subsequent to my reaching the Pole, for none I found later than this date; and these contained nothing scientific, only prayers and despairings. On arriving, therefore, at London, I made straight for *The Times* office, only stopping at a chemist's in Oxford Street for a bottle of antiseptic to hold near my nose, though,

Investigating newspaper reports—

having once left the neighborhood of Paddington, I had hardly any need of this.

So I made my way to the square where the paper was printed, to see that even there the ground was strewn with calpac and pugaree, abayeh and fringed praying-shawl, hob-nail and sandal, lungi and striped silk, all very muddled and mauled; and through the darkling square of the twice-dark pile I passed, to find open the door of an advertisement-office; but, on striking a match, I descried that it had been lit by electricity, and had now to retrace my stumbling steps, till I came to a lamp-shop in an alley, stepping now with care that I might offend no one, for in this enclosed neighborhood I began to undergo tremors, and kept flashing matches, which, so still was the black air, scarcely flickered.

When I got back to the building with a little lighted lamp, I saw a "file" of the paper on a table, and since there were a number of dead there, and I wished to be alone, I took the mass under one arm, the lamp in my other hand, passed behind a counter, and up a stair that led me into a great building and complexity of steps and corridors, where I went peering, the lamp obviously trembling in my hand, for here also were dead. Finally I entered a stately chamber like a board-room, large chairs placed about a table covered with baize, on the table stacks of manuscript permeated with purple dust, and books in book-cases around. This room had been locked upon himself by a single man in a frock-coat, tall, with a pointed grey beard, who at some time had decided to fly from it, for he lay at the door, having dropped dead the moment he opened it; and him, by drawing his boots aside, I removed, locked the door upon myself, sat at the table before the dusty file, and, with the light by my side, began to investigate.

I investigated and read until far into the morning: but God knows . . .

I had not properly filled the little reservoir with oil, so about three in the 'foreday it began to burn sullenly lower, letting sparks, turning the glass grey; and in my heart was the question: "Suppose the lamp goes out before the daylight. . . ."

I knew the Pole and cold, I knew them, but to be frozen by terror. . . . I read, I say, I conned, I would not stop: but I read that night racked by panics such as have never entered into a heart to fancy, my flesh moving and creeping like a pool which, here and there, a breeze breathes on. Sometimes for three, four, minutes the profound interest of what I read would fix my mind, and then I would peruse an entire column, or two, without consciousness of the sense of one phrase, my brain all drawn away to the innumerable troops that camped about me, to musings on the question on whether they might stand, and accuse me: for the worm was the world, and in the air a stirring of cerements, and the taste of the grey of ghosts seemed to infect my throat, and odors of the loathsome tomb my nose, and deep tones of tollings my ears; at the last the lamp smouldered low, low, and my charnel fancy was chockful with the screwing-down coffins, lynch-gates and grave-diggers, and the grating of ropes that lower into the grave, and the first thump of the earth upon the lid of that gaunt and gloomy home of the mortal; that lethal look of cold dead fingers I seemed to see before me, the insipidness of dead tongues, the pout of the drowned, and the vapid froths which ridge their lips, until my flesh was moist as with the stale washing-waters of morgues and mortuaries, and with such sweats as corpses sweat, and the mawkish tear which pauses on dead men's cheeks: for what is one insignificant man in his garment of flesh against mobs and armies of the disembodied, he alone with them, and nowhere another, his peer, to whom to appeal against them? I read, I bent to it: but God knows . . . If a leaf of the paper, which I warily, thievishly, moved, made but one rustle, how did that reveille boom through the haunted halls of my heart, and there was a cough in my swallow which for long I shirked to cough, till it burst with pitiless turbulence from my lips, sending crinkles of cold through my very soul: for with the words which I read were all mixed up visions of hearses crawling, palls, and wails, and crapes, and piercing shrieks of distraction pealing through vaults of catacombs, and all the mournfulness of that valley of shadow, and the tragedy of corruption. Twice during the spectral watches of that night the knowledge that the presence of some mute being brooded at my left elbow so thrilled me, that twice I leapt to my feet to confront it with hairs which bristled in frenzy: after which I must have fainted, for when it was broad day I found my brow dropped upon the papers; and I resolved then never again after sunset to remain in any house: for that night was enough to kill a horse, my God; and that this is a haunted planet I know.

* * *

What I read in *The Times* was not very definite, for how could it be? but in the main it established inferences which I had myself made, and fairly satisfied my mind.

There had been a battle royal in the paper between my collaborator Professor Stanistreet and Dr. Martin Rogers, and never could I have conceived such an indecorous piece of business, men like them calling one another "tyro," "dreamer," and in one place "blockhead." Stanistreet denied that the odor of almonds attributed to the advancing cloud could be due to anything but the excited fancy of the fugitives, because, said he, it was unknown that either Cn, HCn, or K_4FeCn_6 had been given out by volcanoes, and the destruc-

An intelligent article explaining the cloud

tiveness of the cloud could only be owing to CO and CO_2; to which Rogers, in an article characterised by extraordinary acrimony, replied that he could not understand how even a "tyro" (!) in chemical and geological phenomena should rush into print with the statement that HCn had not been given out by volcanoes: that it *had* been, he said, was ascertained, though whether it had been could not affect the question as to whether it was being, since cyanogen, as a matter of fact, was not rare in nature, though not directly occurring, being one of the products of the distillation of pit-coal, and found in roots, peaches, almonds, and many tropical flora; also it had been actually pointed out as probable by more than one thinker that some salt or salts of Cn, the potassic, or the potassic ferrocyanide, or both, must exist in considerable stores at volcanic depths. In reply, Stanistreet in a two-column article used the expression "dreamer," and Rogers, when Berlin had been already silenced, finally replied with his red-hot "blockhead." But, in my opinion, by far the best of the scientific dicta was from the unexpected source of Sloggett, of the Dublin Science and Art Department; he, without fuss, accepted the reports of the fugitives, down to the assertion that the cloud, as it rolled, was mixed from its base to the clouds with tongues of flame, purple, rimmed with rose-color: this, Sloggett explained, being the characteristic flame of both cyanogen and hydrocyanic acid vapor, which, being inflammable, may have become locally ignited in the passage over cities, and only flamed in that limited and languid way because of the ponderous volumes of carbonic anhydride with which they must, of course, be mixed, the dark empurpled color of the cloudmass being due to the presence of scoriæ of the trappean rocks, basalts, green-stone, trachytes, and the various porphyries. This article was remarkable for its discernment, because written so early—not long, in fact, after the cessation of communication with Australia, at which date Sloggett stated that the character of the devastation not only proved an eruption—another, but far greater Krakatoa, doubtless in some South Sea region—but indicated that its most active product must be, not CO, but potassic ferrocyanide (K_4FeCn_6), which, undergoing distillation with the products of sulphur in the heat of eruption, produced hydrocyanic acid (HCn); and this volatile acid, he said, remaining in a vaporous state in all climates above a temperature of 26.5° C., might involve the entire earth, travelling chiefly in a direction contrary to the earth's spin, the only regions which would certainly be exempt being the colder parts of the Arctic circles, where the vapor would condense to the liquid state, and descend as rain. He did not anticipate that vegetation would be deeply affected, unless the event were of inconceivable persistence and activity, for, though the poisonous quality of hydrocyanic acid consisted in its arrest of oxidation, vegetation had two sources of existence—the soil as well as the air; with this exception, all species, down to the lowest forms, would disappear (here was the one point in which he was at fault). For the rest, he fixed the rate of the on-coming cloud at from 100 to 105 miles a day, and the date of eruption as the 14th, 15th, or 16th of April—one, two, or three days after the *Boreal* party reached the Pole; and he ended by saying that, if the facts were as he had stated them, then he could suggest no hiding-place for the race of man, unless such places as mines and tunnels could be made air-tight; nor could even they be of use to any considerable number, except in the event of the lethal state of the air being of brief duration.

* * *

I had thought of mines before, but in a languid way, until this article, and other things that I read, as it were, struck my brain a slap with the notion. For "there," I said, "if anywhere, shall I find a man. . . ."

* * *

I passed out from that building that morning like a man bowed down with age, for the depths of gruesomeness into which I had had glimpses during those hours of gloom made me feeble, my steps tripped, my brain reeled.

I came out into Farringdon Street, and at the Circus, where four streets meet, had under my range of vision four fields of bodies, bodies, clad in a rag-shop of every faded color, or half-clad, or not clad, actually in some cases overlying one another, as I had seen at Reading, but here with a more skeleton appearance: for I saw the swollen-looking shoulders, sharp hips, hollow abdomens, and stiff bony limbs of men dead from famine, the whole having the bizarre air of some *macabre* battlefield of marionettes fallen; and, mixed with them, a multitude of vehicles of all sorts, among which I made my way to a shop in the Strand, where I hoped to find all the information which I required about the excavations of the country; but the shutters were up, and I did not wish to make any noise among these people, though the morning was clear, and it was easy to effect an entrance, for I saw a crowbar on a truck; so I moved on to the British Museum, the cataloguing-system of which I knew, and passed in: no one at the reading-room door now to bid me halt, and in all the round of the reading-room not a soul, except one old man with a bag of goître at his neck, and spectacles, he lying up a book-ladder near the shelves, a "reader" to the last; then, having got at the catalogues, for an hour I was upstairs among the dim sacred galleries of this still place, and at the sight of certain Greek and Coptic papyri, charters, seals, had such a dream of this earth, my good God, as even an angel's pen could not express on paper. Afterwards I went away loaded with half a hundredweight of ordnance-maps which I had stuffed into a bag found in the cloak-room, with three topographical books; then at an instrument-maker's in Holborn got a sextant and theodolite; at a grocer's near the river put into a sack-bag provisions to last me a week or two; and, finding at Blackfriars Bridge wharf-station a sharp white motor-yawl of a few tons, by noon I was cutting my solitary way up the Thames, that flowed as before the Brits were born, and saw it, and built mud-huts there among the forests, and later on the Romans came, and saw it, and called it Tamesis, or Thamesis.

* * *

That midnight, lying asleep on the cabin-cushions of my boat under the lee of an island at Richmond, I had a clear dream, in which something, or someone, came to me, and asked me a question: for it said: "Why do you go seeking another?—that you may fall upon him, and kiss him? or that you may fall upon him, and kill him?" And I muttered sullenly in my dream: "I would not kill him. I do not wish to kill anyone."

* * *

What was essential to me was to know, with definiteness, whether I was alone: for some instinct was beginning to whisper me: "Find that out; be sure, be sure: for without the assurance you can never be—yourself."

I passed into the Midland Canal, and so northward, leisurely advancing, for I was in no sweat, the weather remaining very warm, much of the country still clothed in autumn foliage. I have written, I think, of the terrific recklessness of the tempests witnessed in England since my return: well, the calms were not less intense and novel. This observation was forced upon me: and I could not but be surprised. There seemed no middle course now: if there was a wind, it was a storm; if there was not a storm, no leaf twinkled, not a zephyr fretted the waters. I was reminded of maniacs that laugh now, and rave now—but never smile, and never sigh.

Well, after passing by Leicester on the fourth afternoon, I left my pleasant boat the next morning, carrying maps and compass, and at a small station took engine, bound for Yorkshire, where I loitered away two foolish months, sometimes moving by steam, sometimes by automobile, by bicycle, on foot, till the autumn was quite over.

* * *

There were two houses in London to which I had thought to go, one in Harley Street, one in Hanover Square: but when it came to the point, I would not; and there was an embowered home in Yorkshire, where I was born, to which I thought to go: but I would not, keeping myself for many days to the east half of the county.

One morning, while passing on foot along the coast-wall from Bridlington to Flambro', on looking inward from the sea, I was confronted by a thing which for a moment struck me with profound astonishment—a mansion, surrounded by park, and there at a gate straight before me a board marked: "Trespassers will be Prosecuted." A wild desire—my first—to laugh, to burst with laughter, took me: but I would not, though I could not but marvel at this poor man, with his fantasy that part of a planet was his.

Here the cliffs are some seventy feet high, broken by slips in the upper stratum of clay, and, as I proceeded, mounting always, I encountered gullies in the chalk, down and then up which I had to scramble, till I came to a great mound or barrier, stretching across the promontory, and backed by a ravine, a barrier raised as a rampart apparently by some of those old invading pirate-peoples, who had their hot life-scuffle, and are done now, like the others; then I came to a bay in the cliff, with boats lodged on the slopes, some quite high, though the declivities are steep, and a lime-kiln is there which I explored, but found no one; then, coming out on the other side of the bay, I found the village, with an old tower at one end; and thence, after an hour's rest in the kitchen of a little inn, went out to the coast-guard station, and the lighthouse.

Looking across the sea eastward, the light-keepers here must have seen that cloud of convolving browns and purples, doubtless embroiled with serpents of fire, walking the water, its top in the sky, upon them: for this headland is in the same longitude as London, and, counting from the hour when, as told in *The Times*, the cloud was sighted from Dover over Calais, London and Flambro' must have been overtaken about three on the Sunday afternoon, the 25th of July; and at the view in open daylight of a doom so gloomy—foreknown, but may-be hoped against to the end, and now come—the light-keepers must have fled, if they had not fled before, for here was no one, and in the village few. In this lighthouse, a white tower on the cliff-edge, is a book for visitors to sign their names: and I will write something here, for the secret is between God only and me: After reading some of the names, I wrote my name there. . . .

* * *

The reef before the Head reaches out a quarter-mile, looking bold in the low-water which then was, showing to what extent the sea has pushed back this coast, three weeks impaled on the reef, and a steamer close and huge, waiting for the next movements of the sea, already strewn, to perish. All along the cliff-wall to the bluff crowned by Scarborough

Castle northward appeared those cracks and caverns which had brought me here: so I got down a slope to a rude beach, strewn with blocks of chalk, and never did I feel so paltry and short a being, bays of rock out-flung about me, their bluffs encrusted at the base with crass old leprosies of barnacles and beardedness of seaweed, and, higher up, their whiteness all daubed and time-spoiled, darksome caverns yawning in the enormous withdrawals of the rock-wall. Here, in that morning's walk, I saw three little hermit-crabs, five limpets, and two ninnycocks living their lives in a pool beneath a bearded rock; but what astonished me here, and, indeed, everywhere, in London even, and other towns, was the number of birds that strewed the earth, at some points resembling a rain, birds of almost every sort, including tropic specimens, so that I was compelled to conclude that they, too, had fled before the cloud from country to country, till conquered by weariness and astonishment at Him who by sixty million years of persistence and achievement had completed them into the things they were.

By scrambling over rocks crass with periwinkles, and splashing through sloppy stretches of algæ, which vent a raw stench, I entered one of the gullies, long, winding, its sides polished by the sea-wash, the floor rising inwards, I striking matches in the interior, hearing still from outside the ponderous rushes and jostles of the sea between the rocks of the reef, but now faintly. Here, I knew, I could meet only dead men; but, urged by some curiosity, I searched to the end, wading once through sea-weed three feet deep; but no one there: only belemnites and fossils in the chalk; and after searching several south of the headland, I went northward past it into another bay and place of perched boats, called in the map "North Landing," where, even now, a smell of fish, left by the old crabbers and herring-fishers, was perceptible. Still coves and bays opened as I proceeded, a turf coming down in curves at some parts on the cliff-brows, like wings of hair parted in the middle and plastered on the brow; isolated chalk-masses are common, obelisks, top-heavy columns, bastions; at one point no less than eight headlands stretched to the end of Being before me, each pierced by its arch, Norman or Gothic, in whole or in half; and here again caves, in one of which I found a carpet-bag stuffed with a wet pulp like bread, and, stuck to the rock, a Turkish tarboosh; also, lying in a limestone quarry, five asses: but no man, the east coast having evidently been shunned. Finally, in the afternoon I reached Filey, very tired, and there slept.

* * *

I went onward by steam along the coast to a region of iron-ore, alum, and jet-excavations round Whitby and Middlesborough, and at Kettleness went down to a bay in which is a cave called the Hob-Hole, with excavations all round made by jet-diggers and quarrymen: in the cave a herd of cattle, for what purpose put there I cannot conjecture, and in the jet-excavations I found nothing. Further south is the alum-region, as at Sandsend; but as soon as I saw a works, and the gap in the ground like a crater where the lias is quarried, I concluded that here could have been found no hiding. Then from round Whitby and those rough moors I went on to Darlington, not far now from my home: but I would not continue that way; and, after two days' lounging, started for Richmond and the lead mines about Arkengarth Dale, near Reeth. Here begins a region of mountain, various with glens, fells, screes, scars, swards, becks, passes, villages, river-heads, dales, some of the faces that I saw in it almost seeming to speak to me in a brogue which I knew; but they were not numerous in proportion, for this countryside must have had its population multiplied by hundreds, the villages having rather the air of Danube, Levant, or Spanish villages. In one, named Marrick, the street had become the scene either of a battle or a massacre; and soon I was everywhere coming upon men and women dead from violence: cracked heads, wounds, unhung jaws, broken limbs. But instead of going direct to the mines from Reeth, that waywardness which now governs my mind, as gusts an abandoned boat, took me south-west to the village of Thwaite, which, however, I could not enter, so occupied with dead was every spot. Not far from here I went, on foot now, up a steep road which leads over the Buttertubs Pass into Wensleydale, the day warm and broad, with broad clouds looking like pools of molten-silver which give out grey fumes from their center, throwing moody shades over the dale; and soon, climbing, I could look down upon miles of Swaledale, a panorama of glen and grass, river and cloud-shadow, something of levity being in my step that fair day, for I had left my maps and things at Reeth, to which I meant to return, and the earth, which is very nice, was mine. The ascent was rough, and also long: but, if I paused and looked behind—I saw, I saw. Man's notion of a Paradise reserved for "souls" arose from impressions which the earth made upon his senses, for no seventh heaven can be fairer than this, as his notion of a Hell arose from the mess into which his own baby habits of mentation changed this Paradise: thinking which, I went up into what more and more took-on the character of a mountain-pass, with points of Alpine savagery, heather now on the mountain-sides, a beck sending up its sound, then screes, and scars, a waterfall, a landscape of crags, and lastly a lonesome summit, palpably closer to the clouds.

* * *

Five days later I was at the mines: and here I first saw that wide-spread scene of horror with which I have since become familiar, the story of seven out of ten of them being the same, and brief: selfish "owners," an ousted world, an easy bombardment, and the destruction of all concerned, before the coming of the cloud in many cases. About some of the Durham pit-mouths I have been given the impression that the human race lay collected there, and that the notion of hiding himself in a mine must have entered the head of every man alive, and sent him there.

In these lead mines, as in most vein-mining, there are more shafts than in collieries, and hardly any attempt at artificial ventilation, except at rises, winzes and cul-de-sacs; and I found that, though their depth does not exceed three hundred feet, suffocation must often have anticipated the other death. In nearly every shaft, both up-take and down-take, was a ladder, either of the mine, or of the fugitives; and I was able to descend without difficulty, having dressed myself in a house at the village in a flannel shirt, trousers with circles of leather at the knees, thick boots, and a miner's hat having a socket into which fits a candle; with this and a Davy-lamp, which I carried about for months, I lived for the most part in the depths of the earth, searching for the treasure of a life, to find everywhere, in English duckies and guggs, Pomeranian women in gaudy cloaks, the Walachian, the Mameluk, the Khirgiz, the Bonze, the Imaum, almost every type of man.

* * *

One most brilliant day of Autumn I walked by the market-cross at Barnard, come at last, though with a reluctance in my heart, to where I was born: for I said I would go and see my sister Ada, and the other old one; but I leaned and loitered a long time on the bridge at Barnard, gazing up to the craggy height, heavy with wood that waved, and crowned by the Castle-tower, the Tees round the mountain-base sweeping smooth here and sunlit, but a league down, where I thought of going, brawling bedraggled and lacerated like a sweet strumpet, shallow among rocks under reaches of shadow—the shadow of Rokeby Woods; but I shrank from it, and, instead, went leisurely up the hill-side to the castle, having in my hand a bag with a meal, up the stair in the castle wall to the top, where in my miner's attire I remained three hours, brooding sleepily upon the scene of lush umbrageous wood which marks the way the river takes, from Marwood Chase up above, and where the brabbling Balder bickers in, down to bowery Rokeby daubed now with browns of autumn, the luxury of umbrage lessening away toward the uplands, where there are etherealised reaches of fields, and in the farthest azure remoteness mirages of lonesome moorland. It was not till near three that I went down along the river; then, near Rokeby, up the old hill: and there, as of old, was the little black square with yellow letters on the gate-wall:

No house, I believe, of this countryside was empty of invaders, and they were in Hunt Hill, too—three to the right of the garden-patch, where the lilac, among weeds now, had once grown from rollered grass; and in the bush-wilderness to the left, which had always been wilderness, one more; and in the breakfast-room three; and in the new clinker-built attachment two, half under the billiard-table; and in her room overlooking the porch the long form of my mother on her bed, her left temple battered in; and beside the bed, face down on the planks, black-haired Ada in a night-dress.

Of the men and women who died they two alone had burying, for I delved a hole with the stable-spade beneath the cedar, and wound them in sheets for shrouds, feet and form and countenance, and, not without throes and qualms, bore and buried them there.

* * *

Some time passed after this before the multitudinous and perplexing task of visiting the mine-regions anew claimed me, I meantime finding myself at a place named Ingleborough, which is a table-mountain with a summit of twenty acres, from which the sea is visible across Lancashire to the west; and in the flanks of this strange mount are a number of caves which I scrutinized during three days, sleeping in a tool-shed at a very rural and flower-embowered village, for every room in it was crowded, a place marked Clapham in the chart, in Clapdale, which latter is a dale penetrating the slopes of the mountain: and there I found by far the vastest of the caves which I found, having climbed a path from the village to an arch, screened by trees, leading into the limestone cliff; nor had I proceeded three yards ere I saw the traces of a battle here—all this region, in fact, had been invaded, for the cave must have been famous, and for some miles round it the dead were numerous, so that the approach to the cave was a case for care, if the foot was to be saved from pollution. There had always been an iron gate across near the entrance, within which a wall had recently been built across, shutting in I do not know how many, and both gate and wall had been stormed, for there still lay the sledges which had done it. So, having a lamp, and on my hat the candle, I went on quickly, seeing it useless now to choose my steps where there was little to choose, through a passage incrusted with a scabrous petrified lichen, the roof low, covered with down-looking cones like a forest of children's toy-trees. I then came to a hole in a curtain of stalagmitic formation opening into a cavern which was quite animated and fistal with flashes, sparkles, diamond-lusters, hung in their myriads upon a movement of the eye, produced by wet stalagmites, down the center of which ran a lane of clothes and hats and faces: over which with hasty reluctant foot I somehow trod; the cavern all the time widening, staliactites on the roof of every size, from cow's breast to titan's club; and now everywhere the wet drip, drip, as it were a crowded bazaar of sweating brows and ardent steps, in which the one business is to drip. Where stalactite meets stalagmite there are pillars; where stalactite meets stalactite there are elegances, draperies, delicate fantasies; there were also ponds in which hung heads and feet; and there were regions where the roof, which continually reared itself, was reflected in the chill sheen of the floor. Suddenly down came the roof, the floor went up, and they seemed to meet before me; but I found an opening, through which, rowing myself on the belly over slush in repulsive proximity to dead personalities, I issued out to a floor of sand under a tunnel which is arched and narrow, grum and dull, without stalactites, in a mood of monks and catacombs and the route to the tomb: the dead there fewer, proving that the general mob had not had time to penetrate so far inward, or else that those within had gone out to defend, or to hearken to the storm of, their citadel. This passage brought me to a hall, the amplest of all, loftily vaulted, fraught with genie riches and buried treasures of brilliance, the million-fold *ensemble* of flashes dancing schottishe with the eye, as it shifted or was quiet, this place being quite half a mile from the entrance; and here my prying light could find only nineteen dead, and at the remote end two holes in the floor, just big enough to admit the body, through which from below arose a noising of falling water; both of which holes, I could see, had been filled-in with cement-concrete—wisely, I think, for a current of air from somewhere seemed to be breathing through them, and must have resulted fatally; but both of the fillings had been broken through—by the ignorant, I guess, who thought to get to a den yet beyond. I had my ear an hour at the larger of these holes, hearkening to the charm of that chanting down below in the dark; and afterwards, goaded by my desire to be thorough, got a number of robes from the bodies, tied them together, then tied one end round a pillar, and having put my mouth to the hole, calling "*Anyone? Anyone?*", let myself down by the rope of robes, the candle-light at my brow; but I had not descended far down those mournful darks when my left foot dipped into liquid, and instantly the feeling of appalment pierced me that all the evil things in Gehenna were at my leg to get me down to Hell: and I was up quicker than I went down; nor did my flight rest until, with a sigh of deliverance, I found myself out in the open.

* * *

After this, seeing that the autumn warmth was passing away, I set myself with more system to my task, and during six months stuck to it with steadfast will and strenuous assiduity, seeking, not indeed for a man in a mine, but for some evidence of the possibility that a man might be alive, visiting in that time Northumberland and Durham, Fife and Kinross, South Wales and Monmouthshire, of the Isle of Man, Waterford, Down; I have gone down the 360-ft. ladder of the graphite-mine at Barrowdale in Cumberland half-way up a mountain 2,000 feet high; and have visited where cobalt and manganese are mined in pockets at the Foel Hiraeddog mine in Flintshire, and the lead and copper workings in Galloway; the Bristol coal-fields, and the mines of South Staffordshire, where, as in Somerset, the veins are thin, and the mining-system is the "long-wall," whereas in the North the system is the "pillar-and-stall;" I have visited the open workings for iron-ores of Northamptonshire, and the underground stone-quarries, and the underground slate-quarries in the Festiniog district of North Wales; also the rock-salt workings; the tin, copper and cobalt workings of Cornwall; and where the minerals were brought to the surface on the backs of men; and where they were brought by adit-levels provided with railroads; and where, as in old Cornish mines, there are two ladders in the shaft, moved up and down alternately, see-saw, and by skipping from one to the other at right moments you ascended or descended; the Tisbury quarries in Wiltshire, the Spinkwell in Yorkshire; and every tunnel, and every recorded hole: for something urged within me, saying: "You must be *sure* first, or you can never be—yourself."

* * *

At the Farnbrook Coal-field, in the Red Colt Pit, my inexperience nearly ended me: for though I had a theoretical knowledge of all British workings, I was, in my practical relation to them, like a man who has learned seamanship on shore. Here I arrived on the 19th of December to find the dead accumulated beyond precedent, the plain being as strewn as a reaped field with stacks, and near the bank more strewn, filling the only house within sight of the pit-mouth—the houselet provided for the company's officials—even lying over the mountain-heap of "wark," composed of the shale and *débris* of the working; and here I did not, as usual, see any rope-ladder fixed by the fugitives in the ventilating-shaft (which, usually, is not deep, being also the pumping-shaft, containing a plug-rod at one end of the beam-engine which works the pumps); though, on looking down the shaft, I discerned clothes and a rope-ladder, which a group of the fugitives, by hanging their weights to it, must have dragged down, to prevent the descent of yet others: so my only way of going down was by the pit-mouth, and after some hesitation I decided, very rashly, first providing for my coming up again by getting a coil of half-inch rope from the bailiff's

Examining a coal mine

office, rope at most mines being so profuse, that it seemed as if each fugitive had provided himself in that way; and this rope I threw over the beam of the beam-engine with both ends at the bottom of the ventilating-shaft: in this way I could come up by tying one rope-end to the rope-ladder down there, hoisting the ladder, fastening the other rope-end below, and climbing the ladder; and now, to go down, I lit the pit-mouth engine-fire, started the engine, and brought up the cage from the bottom, the 300 yards of wire-rope winding with a quaint deliberateness round the drum, reminding me of a camel's nonchalant leisurely obedience: so when the four meeting chains of the cage appeared, I stopped the ascent, tied a string to the knock-off gear, carried its other end to the cage, in which I had five companions, lighted my hat-candle, which was my test for choke-damp, and the Davy, and without reflection pulled the string. First the cage gave a little up-leap, then began to descend, normally, I thought—though the candle at once went out—nor had I the least fear, for though a draught blew up the shaft, that happens in shafts; *this* draught, however, soon got to be too vehemently boisterous, I saw the lamp-light struggle, the dead cheeks shiver, heard the cage-shoes go singing down the guides, and quicker we went in that facile descent of Avernus, slipping light, then raging, a rain of sparks shooting from the shoes and guides, a gale in my brain and eyes and breath. When we bumped upon the "dogs" at the bottom, I was tossed a foot upwards with those stern-faced others, then lay among them as one of them.

It was only when, an hour later, I sat disgustedly reflecting on this mauling, that I remembered that there used always to be some "hand-working" of the engine during the cage-descents, an engineman reversing the action by a handle, to prevent bumping. However, the only permanent hurt was to the lamp, and I found thousands in the workings.

I then got out into the coal-hole, a hall 70 feet square, the floor paved with iron sheets, some holes round the wall, dug for some purpose which I never could discover, wagons full of coal and shale standing about, and all among the wagons, on them, under them, bodies, clothes. I got a new lamp, pouring in my own oil, and went down a ducky-road, very rough, with rollers over which ran a rope to the pit-mouth for drawing up the wagons; and in the sides here, at regular intervals, man-holes, within which to rescue one's self from wagons tearing down; and within these man-holes here and there a dead, in others things to eat, and at one place a dead heap, the air here hot at 65 degrees, and getting hotter with the descent.

This ducky led me down into a standing—a space with a turn-table—which I made my base of operations, here being a number of putts like punts on carriages, with wagons, such as took the coal from putt to pit-mouth; and, raying out from this standing, avenues, some ascending as guggs, some descending as dipples, and the dead here all arranged in groups, the heads of this group pointing up this gugg, of that group down that dipple, the central space, where weighing was done, nearly empty: and the dumbness of this deep place among all these multitudes I found extremely gravitating and hypnotic, dragging me also into their passion of dumbness, in which they lay, all, all, so fixed and veteran; and at one period I fell a-staring, nearer perhaps to death and the inane Gulf than I knew; but I said I would be strong, and not sink into their habit of stillness, but let them keep to their own way, and follow their own fashion, and I would keep to my own way, and follow my own fashion, nor yield to them, though I was but one against many: so I pulled myself together, and, getting to work, holding on to the drum-chain of a gugg, I got up, stooping under a roof three feet high, until I came upon the scene of another battle: for in this gugg nineteen of the mine-hands had clubbed to wall themselves in, and I saw them lie there behind their stormed wall with their bare feet, trousers, but naked bodies, countenances all fierce and wild, their grime streaked with sweat-furrows, candles in their hats; and, outside, their own "getting" mattocks and boring-irons to besiege them. Thence I went along a curving twin-way, into which, every thirty yards or so, opened one of those putt-ways called topples; and all about here, in twin-way and topples, were ends and corners, not one of which had been left without its walling-in, and only one was now intact, some, I fancied, having been broken open by their own builders at the goad of suffocation or hunger. The one intact I broke into with a mattock—it was only a cake of plaster, but air-tight—and in a space not nine feet long behind I found the foul corpse of a carting-boy, with guss and tugger at his feet, and the pad which protected his head in pushing the putts, and loaves, sardines, bottled beer, and five or six mice which pitched shrieking through the opening which I made, shocking me, there being of dead mice extraordinary swarms in all this mine-region. I went back then to the standing, and at one point where there was a windlass and chain lowered myself down a "cut"—a pit sunk to a lower coal-stratum. Down there, fancying I could hear the perpetual rat-rat of notice once exchanged between the putt-boys below and the windlass-boys above, I proceeded down a dipple to another standing, for in this mine there were six, perhaps seven, veins: and there I came upon the acme of the drama of this Tartarus, all here being not merely thronged, but at some points a congestion of flesh, reeking a smell of peach mixed with the stale coal-odor of the pit, for here ventilation must have been limited; and masses here had been mown down by only three hands, as I found: for through three holes in a wall built across a gugg stuck out a little three muzzles, plugged in the plaster, which must have glutted their guts with massacre; and when, after a horror of disgust at wading through a dead sea, I got to the wall and peeped through a hole, I made out a man, two youths, two women, three girls, and heaps of cartridges and provisions, the hole having no doubt been pierced from within at the point of suffocation, when the poison must have entered; and I conjectured that here must be the mine-owner, director, or manager, with his family. In another dipple-region, when I had re-ascended to a higher level, I nearly fainted before I could retire from the commencement of a region of after-damp, where there had been an explosion, the corpses now all bald, ravished and scaramouch. But I did not desist from searching every other district: no momentary work, for not till six did I go up by the pumping-shaft rope-ladder.

* * *

One day, standing in that region of rock and sea called Cornwall Point, whence one can watch the postillion rocks of Land's End dash out into the sea, and the flash of all the wild white steeds of the sea between, and not a building in sight, on that day I finished what I may name my official inquisition.

In going away from that place, walking northward, I came upon a house by the sea, a beautiful house of bungalow type with a sea-side expression, its special feature a spacious loggia or verandah, sheltered by the overhanging of the upper story, the exterior of rough-hewn blocks with a batter, the roofs of low pitch, covered with green slates, a feeling of strength and repose heightened by the long horizontal lines, at one end of the loggia a turret containing a study or nook; and in this place I lived three weeks. It was the house of the poet Machen, whose name, as soon as I saw it, I remembered well, and he had married a beauty of eighteen, obviously Spanish, who lay on the bed in the large bright bedroom to the right of the loggia, on her left breast being a baby with an india-rubber comforter in its mouth, both mother and child wonderfully preserved, she still quite lovely, white brow under curves of raven hair. The poet, though, had not died with them, but was in the room behind in a loose silky-grey jacket, at his desk—writing a poem! writing, I could see, wildly quick, the place littered with the written leaves—at three o'clock in the morning, when, as I knew, the cloud overtook this end of Cornwall, and stopped him, and put his head to rest on the desk; and the little wife must have got sleepy, waiting for it to arrive, probably sleepless for nights previously, and gone to bed, he perhaps promising to follow to die with her, but bent upon finishing his poem, writing feverishly on, running a race with the cloud, thinking, no doubt, "just two couplets more," until the thing came, and put his head on the desk; and I do not know what I ever encountered anything so complimentary to my race as this Machen, and his race with the cloud: for it is clear now that the better kind of those poet men did not write to please the dim inferior tribes who might read them, but to deliver themselves of the divine warmth that swarmed within their breast, and, if all the readers had been dead, still they'd have written, and for God to read they wrote. At any rate, I was so pleased with these poor people, that I stayed with them three weeks, sleeping on a couch in the drawing-room, a place rich in lovely pictures and faded flowers, like all the house: for I would not touch the young mother to remove her. And, finding on Machen's desk a note-book with soft covers, dappled red and yellow, I took it, and in the little turret-nook wrote day after day for hours this account of what has happened, and I think I may continue to write, for I find in it a comfort and company.

* * *

In the Severn Valley, somewhere in the plain between Gloucester and Cheltenham, in a rather lonely spot, I at that time travelling on a motor-bicycle, I spied a curious erection, went to it, and found it perhaps fifty feet square, made of brick, the flat roof, too, of brick, and not one window, and only one door, which I found open, rimmed with india-rubber, airtight when closed. Inside I came upon fifteen English people of the dressed class, except two, who were bricklayers: six ladies, nine men; and, farther within, two more, men, who had their throats cut, whether through sacrificing themselves for the others when breathing difficulties commenced, or killed by the others, was not clear: along one wall provisions; and a chest full of oxide of manganese, with an apparatus for producing oxygen—a foolish thing, for additional oxygen could not alter the quantity of carbonic anhydride breathed out, this being a narcotic poison; and finally they must have opened the door, and so met their death. I believe that this erection was run up by their own hands under the direction of the two bricklayers, for they could not, I suppose, have got workmen, except on the condition of the workmen's admission: on which condition they would employ as few as possible.

In general, I observed that the rich must have been more urgent and earnest in seeking escape than the others: for the poor realized only the near and visible, lived in today, and cherished the notion that tomorrow would be the model of today. In an out-patients' waiting-room, for instance, in the Gloucester infirmary, I chanced to see an astonishing thing: four old women in shawls, come to have their ailments medicined on the day of doom; and these, I concluded, had been unable to realize that anything would occur to the daily old earth which they knew and had footed with assurance on: for, if everyone was to perish, they must have felt, who would preach in the Cathedral on Sunday evenings? In an adjoining chamber sat an old doctor at a table, his stethoscope-tips still clinging in his ears, a woman with bared bosom before him; and I said to myself: "Well, this old man, too, died doing his work. . . ."

In one surgical ward of this infirmary the patients had died, not of the poison, nor of suffocation, but of hunger—the doctors, or someone, having made the ward air-tight, locking them in, for I came upon a heap of maimed shapes, mere skeletons, crowded round the door within; and I knew that their death was not due to the cloud-poison, for the pestilence of the ward was uninformed with that almond charm which did not fail to have embalming effects upon the bodies which it saturated: so that I rushed from that place; and, thinking it a pity and a danger that such a pest should be, I set to work to collect things to burn the building.

It was while I was seated in an easy-chair in the street the following evening, smoking, watching the combustion of this structure, that something was suddenly born in me, something out of Hell, and I smiled a smile that never man smiled. And I said: "I will burn: I will return to London. . . ."

* * *

On this Eastward journey, stopping for the night at Swindon, I had a dream: for I dreamed that a little old man, brown, bald, with a bowed back, whose beard ran in one streamlet of silver from his chin to reach out over the floor, said to me: "You think that you are alone on the earth, its despot; well, have your fling; but as sure as God lives, as God lives, as God lives"—six times—"sooner or later, later or sooner, you will meet another. . . ."

And I started from that slumber with the brow of a corpse, wet with sweat. . . .

* * *

I returned to London on the 29th of March, arriving within a hundred yards of the Northern Station one windy dark evening near eight, where I alighted to walk to Euston Road, then eastward along it, till I came to a shop which I knew to be a jeweller's, though it was too dark to discern any painted words. The door, to my annoyance, being locked, like almost all the shop-doors in London, I went looking about for something heavy, found a laborer, cut one boot from the shrivelled foot, and beat at the glass till it came raining, then entered.

No horrors now at that clatter of glass; no sick qualms; my pulse steady; my head high; my step royal; my eye cold.

* * *

I was going to a hotel, and was not sure of finding sufficient candlesticks, for I had acquired the habit of sleeping with at least sixty about me; and their pattern, age, material, was of importance to me: so I selected from that shop ten of ecclesiastical brass, then found a bicycle, pumped it, tied my bundle to it, and set off; but I had not gone ten jolted yards, when a fork snapped, and, on finding myself across the knees of a Highland soldier, I flew with a shower of kicks upon the foolish thing: and this was my last attempt in that way in London, the streets being in an unsuitable condition.

Throughout that gloomy night it blew great guns: and during nearly three weeks, until London was no more, there was a booming of winds that seemed to bemoan her doom.

* * *

I slept in a Bloomsbury hotel, and, waking the next day at ten, ate with shiverings in the banqueting-hall, went out then, and, beneath drear skies flying low, walked all the way to the West District, accompanied by a prattle of flapping flags—fluttering robes and rags—and grotesque glimpses of decay. I was warmly clad, but the bizarrerie of the European clothes which I wore had become an offence and mockery in my eyes, so at the first moment I set out whither I knew that I should discover such clothes as a man might wear: to the Turkish Embassy in Bryanston Square.

I had been acquainted with Redouza Pasha, but could not recognize him here in an invasion of hanums in their veils, fierce-looking Caucasians in skins of beasts, a Sheik-ul-Islam in his green cloak, three emirs in cashmere turbans, two tziganes, their brown mortality more abominable still than the Western's; but upstairs I soon came to a boudoir odorous of that reclusion and dim mystery of Orient homes: a door encrusted with mother-of-pearl, sculptured roof, candles clustered in tulips and roses of opal, a brazen brasero, and, all in disarray, the silken chemise, the winter-cafetan doubled with furs, cabinets, sachets of aromas, babooshes, stuffs. When, after two hours, I went from the house, I was bathed, anointed, combed, scented, robed.

* * *

I have said to myself: "I will ravage and riot in my kingdoms, I will rage like the Cæsars, and be a withering blight where I pass like Sennacherib, and wallow in soft delights like Sardanapalus; I will raise me a palace wherein to stroll and parade my monarchy before the Gods, its stones of gold, with rough frontispiece of ruby, and cupola of opal, and porticos of topaz: for there were many men to the eye, but there was One only, really: and I was he." And always I knew it—some whisper which whispered me: "*You* are the Arch-one, the motive of the world, Adam, and the rest of men not much." And they are gone—all! all!—as no doubt they merited: and

I, as was meet, remain. And there are wines, opiums, haschish; and there are oils and spices, fruits and oysters, and soft Cyclades, luxurious Orients. I will be restless and dreadful in my territories; and again, I will be languishing and fond. I will say to my soul: "Be full."

* * *

I watch my mind, as in that old time I used to watch a precipitate in a test-tube, to see into what sediment it would settle.

I am very averse to work of any sort, so that the necessity for performing the simplest little labors will rouse me to indignation; but if a thing will contribute greatly to my ever-growing voluptuousness, I will undergo a considerable amount of drudgery to accomplish it, though without steady effort, being liable to side-winds and whims, and wayward relaxations.

In the country I became pretty irritable at the necessity which confronted me of sometimes cooking some vegetable—the only food which I was forced to take some trouble over, for meats and fish, some delicious, I find already prepared in guises which will remain good centuries after my death, should I ever die; in Gloucester, however, I found peas, asparagus, olives, and other greens, already prepared to be eaten without base cares, and these, I now see, exist everywhere in stores that may be named boundless: so I now take my repasts without more cark than when a man had to carve his fowl, though that mote I sometimes find tiresome. There remains the degradation of lighting fires for warmth, for the fire at the hotel always goes out while I sleep; but that is an inconvenience only to this zone, to which I shall soon say farewell.

During the afternoon of my second day in London I sought out a strong motor in Holborn, oiled it a little, set off over Blackfriars Bridge, making for Woolwich through that more putrid London of the south; and one after the other I connected eight drays and cabs to my motor behind, having cut away the withered horses, using the reins, &c., as couplings: and with this train I rumbled eastward.

Half-way to Woolwich I happened to look at my old silver chronometer of *Boreal*-days—and how I can be rushed into these agitations by a nothing, a *nothing*, my good God, I do not know! just by the fact that the hands chanced to point to 3.10, the moment at which the clocks of London stopped—for each town has its thousand weird fore-fingers, pointing, pointing still, to the moment of doom—3.10 on a Sunday afternoon in London. I first observed it in going up the river on the face of that "Big Ben," and now find that they all, all, have this 3.10 mania, time-keepers still—of the end of Time; noting for ever more that one moment: for the cloud-mass of powdery *scoriæ* must have instantly stopped their escapement, and they had fallen silent with man; but in their insistence upon this particular minute I had found something so solemn, yet mock-seldom, ironic, and as it were addressed to *me,* that when my own watch dared to point to the same moment, I was thrown into one of those panting paroxysms, half rage, half horror, which have hardly harrowed me since I abandoned the *Boreal.* On the morrow, alas, another was in store for me; and once more on the morrow after.

* * *

My train was so execrably slow, that not until five did I arrive at the Woolwich Royal Arsenal, and, as it was then too late to work, I uncoupled the motor, and turned back for London; but, overcome by languor, I got candles, stopped at the Greenwich Observatory, and within that gloomy pile burned my watch-lights for the night, musing upon the tempest belling. But, astir early, I was back by ten at the Arsenal, and started to analyze some of that vast and multiple entity. Parts of it seemed to have been abandoned in undisciplined haste, and in the Cap Factory, which I first entered, I found tools by which to effect an entry into any part, my first search being for time-fuses, of which I required some thousands, and after a hunt found a host arranged in rows in a range of buildings called the Ordnance Store Department. I then descended, walked back to the wharf, brought up my train, and began to lower the fuses in bagfulls by ropes through a chute, letting go each rope as the fuses reached the dray. However, on winding one fuse, I found that the mechanism would not go, choked with scoriæ; and I had to resign myself to the task of opening and dusting every one: a wretched labor in which I spent that day like a laborer till about four, when I threw them to the devil, having done two hundred; then hummed back in the motor to London.

* * *

That same evening as it was becoming dark I paid a visit to my old self in Harley Street, a bleak tempest that hooted like whooping-cough sweeping the streets: and at once I saw that even I had been invaded, for my door swung open, banging, a catch preventing it from slamming; and in the passage my car-lamp shewed a young man who seemed a Jew, seated as if in sleep with dropped forehead, a silk-hat, tilted back, pressed down upon his head to the ears; and, lying, six more, a girl with Arlesienne head-dress, a negress, a Deal lifeboat's-man, and three of uncertain race; the first room—the waiting-room—still more numerously occupied, though there still on the table lies the volume of *Punch,* the *Gentlewoman,* and the book of London views in heliograph. Behind this, descending the two steps to the study and consulting-room, I found as ever the revolving-top desk, but on my little shabby-red sofa a large lady too big for it in shimmering grey silk, round her left wrist a *trousseau* of gold trinkets, her head dropped right back, almost severed by an infernal gash from the throat. Here were two old-silver candlesticks, which I lit, went upstairs, and in the drawing-room sat my old housekeeper, placidly dead in a rocking-chair, her left hand pressing down a batch of the piano-keys, among many strangers. But she was very good, had locked my bed-room against intrusion, and, as the door stands across a corner behind a green-baize curtain, it had not been seen, at least not forced. I found the key hung on the switch by the door: and there lay my bed intact, and everything tidy.

But what interested me in that room was the thing on the wall between wardrobe and dressing-table—that gilt frame—and that man painted within it there: myself in oils, done by—no, I forget his name now, towering celebrity he was; in a studio in St. John's Wood, I remember, he did it, and people said that it was quite a work of art. I suppose I was standing before it thirty minutes that night, holding up the bits of candle, lost in wonder, in amused contempt at that thing there. It is I, certainly, that I must admit, the high-curving brow—really a King's brow, after all, it strikes me now—and that vacillating look about the eyes, and the mouth which used to make my sister Ada say "Adam is weak and luxurious." Yes, that is wonderfully done, the eyes, that dear, vacillating look of mine: for although it is rather a staring look, yet one can almost see the pupils stir from side to side: very well done. And the longish face; and the rather thin, stuck-out moustache, shewing both lips which pout a bit; and the hair, nearly black; and the rather visible paunch; and, oh, Heaven, the neat cravat—ah, it must have been *that—the cravat*—that made me burst into laughter! "Adam Jeffson," I muttered when it was over, "could that thing in the frame have been *you*?"

I cannot quite state why the tendency toward Orientalism—Oriental dress—all the manner of an Oriental monarch—has taken full possession of me, but so it is: for surely I am hardly any longer a Western, "modern" mind, but a primitive, Eastern one. Certainly, that cravat in the frame has receded a million leagues, ten thousand forgotten æons, from me! Whether this is a result of my own personality, of old acquainted with Eastern notions, or whether, perhaps, it is the natural accident to any soul emancipated from trammels, I do not know; but I seem to have gone right back to the beginnings, to resemblance with man in his first, simple, gaudy conditions: my hair, as I sit here, already hanging an oiled string down my back; my beard sweeping scented in two opening whisks to my ribs; I have-on the *izar*, a pair of drawers of yomani cloth like cotton, with yellow stripes; over this a shirt, or *quamis,* of white silk, reaching to my calves; over this a vest of crimson, gold-embroidered, the *sudeyree;* over this a khaftan of silk, green-striped, reaching to the ankles, encircled at the waist with a gaudy shawl of Cashmere for girdle; over this a wide-flowing torrent of white drapery, warm, lined with ermine; on my head the skull-cap, covered by a high cap, scarlet with blue tassel; on my feet blue-morocco shoes covered over by thick crimson-morocco babooshes. My ankles—my ten fingers—my wrists—are heavy with gold and silver ornaments; and in my ears, which, with considerable pain, I bored three days since, are two needle-splinters, to prepare the holes for rings.

* * *

O Liberty! I am free. . . .

* * *

While I was going to visit my home in Harley Street that night, at the moment when I turned north from Oxford Street, this thought, hissed into my ear, was all at once seething in me: "If now I should lift my eyes, and see a man walking yonder—*at the corner there*—turning out of Harewood Place, what, my good God, should I do?" and I turned my eyes, leering suspicious eyes, furtively turned, and I peered deeply with lowering brows.

Horribly frequent has this nonsense become with me—in streets—in nooks of the country: the assurance that, if I but glance just *there,* I shall see—*must* see—a man; and glance I must, though I perish; and when I glance, though each hair creeps and rears, yet in my glare, I feel, is monarch indignation, my neck sticks lofty as sovereignty itself, and on my forehead sits all the lordliness of Persepolis and Serapis.

To what point of wantonness this awfulness of royalty may lead me, I do not know. I will watch, and see. It is written, "It is not good for man to be alone;" but, good or no, the arrangement of one-planet-one-inhabitant already seems to me, not merely a natural, but the *only* natural and proper, condition: so much so, that any other arrangement has now, to my mind, a kind of unlikely, wild and far-fetched unreality, like the utopian schemes of dreamers and faddists. That the earth should have been turned out for *me*—that London should have been erected in order that *I* might enjoy the heroic spectacle of its destruction—that history should have existed to accumulate for *my* pleasures its inventions, its stores of wine and spice—no more extraordinary does it all seem to me than to some little duke of the old scheme of things seemed the "owning" of fields of which his forefathers slew the holders; but what strikes me with some surprise is that the new scheme should have come to seem *so* commonplace and natural—in nine months. The mind of Adam Jeffson is adaptable.

* * *

I sat a long time thinking such things by my bed that night, till finally I was inclined to sleep there; and, lacking candlesticks, I remembered that Peter Peters, three doors away on the other side, had had five candelabra in his drawing-room; so I said to myself: "I will search for candles in the kitchen, and, if I find any, will go and get Peter Peters" candelabra, and sleep here."

I took then the two lights which I had, my good God, went down to the basement, and there found three packets of candles, the fact being that the cessation of gas-lighting had compelled everyone to provide himself in this way, for there were many everywhere. With these I reascended, went into the little alcove where I had kept some drugs, got a bottle of carbolic oil, and went dashing all the corpses; then left the two candles on the waiting-room table, and, with the study-lamp, passed to the front-door, which was irascibly banging. I stepped out to find the tempest heightened to a mighty turbulence (though it was dry), which instantly snatched at my clothes to whirl them into a flapping cloud about me, and my lamp was out. I persisted, however, half blinded, to Peters' door, found it locked, though near by was a window, the sash up, into which with little difficulty I lifted myself; but my foot, as I lowered it, stood on a body, and this fretted and upset me, so that I hissed a curse, and passed on scraping the carpet with my soles, that I might hurt no one: for I did not wish to hurt anyone. The murk here was not deep, I could recognize Peters' furniture, but when I passed out into the passage all was blackness, and I, depending upon the lamp, had left the matches in the other house. Still, I felt my way to the stair, my foot was on the lowest step, when I was arrested by a shaking of the front-door, which someone seemed to be at with hustlings and the most urgent poundings, while I stood with stern brows, peering, two or three minutes: for I knew that, if I once yielded to the flinching at my heart, no mercy would be shown me in this house of tragedy, but thrilling shrieks would of themselves emanate, to ring through its haunted halls; and, though the rattling continued an inordinate interval—insistent, imperative—so that I thought it could hardly fail to force the door, I whispered to my heart that it could only be the winds struggling at it as with the vigor of a wrist; and presently I groped on up by the rail—in my brain now the remembrance of a dream which I had once dreamed in the *Boreal* of the woman Clodagh, how she left drip a fluid like pomegranate-seeds into gruel, and tendered it urgently to Peter Peters, and it was a dreadful purging-draught; but I would not stop, step by step went up, though I suffered, my brows peering at the deep dark, my heart shocked at its own rashness, till I got to the first landing; but, as I turned there to ascend the second part of the stair, my left palm touched something deathly cold; on which, making some movement of terror, my foot struck something, and I stumbled over a table there: a horrible row followed, for something dropped to the ground; and in that moment, ah, I heard—a voice—human—that uttered words—the voice of Clodagh, for I knew it; yet not the voice of Clodagh in the flesh, but clogged with clay and worms, choked with effort, and thick-tongued; and in that grisly croaking of the grave I heard the words:

"Things being as they are in the matter of the death of Peter . . ."

There it stopped dead, leaving me so sick, my God, so sick, that I could hardly gather my garments round me to fly, fly, fly, soft-footed, whimpering in pain, down the steps, down like a sneaking thief, but quick, snatching myself away, then wrestling with the catch of the door, which she would not let me open, conscious of her all the time behind me, watching me. And when I did get out, I was away up the length of the street, trailing my *jubbah,* glancing backward, gasping, for I thought that she might dare to follow, with her daring will; and all that night I lay on a bench within the wind-tossed and darkling Park.

* * *

The first thing which I did when the sun was up was to return to that place: and I returned with a hard and masterful will.

Approaching Peters' house I remarked now, what the dark had concealed from me, that on his balcony was someone, alone there—a slight iron structure, connected to its roof by three voluted pillars: at the middle one of which was a woman —kneeling—her arms clasped about the pillar, her face uplooking; and never did I see aught more horrid: the curves of the woman's bust and hips still well-preserved in a dress of red, much faded now; her reddish hair floating loose in a cloud about her; but her face in that exposed place had been eaten away by breeze and gale to a noseless skeleton which grinned from ear to ear, the jaw dropped—horrid in contrast with the grace of body, and frame of hair. I meditated upon her long that morning from the opposite pavement: the locket at your throat contained, I knew, my portrait, Clodagh, poisoner . . .

I thought that I would go into that house, and walk through it from top to bottom, and sit in it, and spit in it, and stamp in it, in spite of anyone: for the sun was now high. So I went in and up the stair to the place where I had been perturbed, and had heard the words: and here a rage took me, for I understood that I had been made the dupe of the malignant wills that beset me, and the laughing-stock of Those for whom I care not a fig, seeing that from a little table there I had knocked to the ground in my stumble a phonograph with a horn, which I now kicked down the stair: for I gathered that its clock-work, stopped up by the scoriæ, had been jogged into a few movements by the shock of the fall, causing it to gossip those thirteen words to me and stop; and I was indignant then, but have since been glad: for I was thereby given the idea of gathering "records," and have been touched to strange sensations, sometimes thrilled, at listening to this stillness of Eternity being disturbed by those voices sounding upon me out of the void.

* * *

Well, most of that same day I spent in a chamber at Woolwich, dusting out, sometimes oiling, time-fuses: a job in which I acquired such ease; that each finally kept me just ninety seconds, so that by evening I had done 500, these little things being pretty simple, easily made, most containing a tiny dry-cell which sparks at the running-down moment, while others ignite by striking. I arranged them in rows in the van, and passed the night in an inn near the Barracks, having brought candle sticks from London; and I so arranged the furniture round the bed as to get an altar of candles mixed with vases containing palms, amid which I scattered a fragrance of ambergris from some Arab sachets which I had, and in the bed a bottle of sweet Chypre-wine, with bonbons, nuts, and havannas; and, lying there, I meditated with a smile which I knew to be malign upon that lust in me which was urging me through all those drudgeries at the Arsenal, I who shirked all work as unroyal. So, however, it was: and the next morning I was at it again, my fingers stiff with cold, for the gale blew keen; but before noon I had 800 fuses, and, judging these sufficient to begin with, got into the motor, and took it round to a place called the East Laboratory, a series of buildings, where I knew that I should find whatever I wanted: and I prepared my mind for a day's labor. In this place I found stores on stores: mountains of percussion-caps, more chambers of fuses, small-arm cartridges, shells, and all those murderous chemicals, amaking and made, with which man exterminated himself: clever, and yet . . . Queer mixed people, like ægipeds, and mermaids, and absurd immature births. At any rate, their lyddites, melanites, cordites, galignites, dynamites, toluols, powders, jellies, oils, marls, came in very well for their own destruction: for by three o'clock I had so worked, that I had on the first vehicles the phalanx of fuses, with kegs and cartridge-boxes full of powder, of explosive cottons and gelatines, liquid nitro-glycerine, earthy dynamite, with bombs, reels of cordite, two pieces of tarred cloth, an iron ladle, a spade, a crow-bar; then the cabs containing coal and cans of oil. And first, in the Laboratory, I connected a fuse with a huge tin of blasting-gelatine, and I timed the fuse for the midnight of the seventh day thence; after which I visited the Carriage Department, the Ordnance Store Department, the Powder Magazines in the Marshes, traversing, it seemed to me, miles of building; and in some I laid coal-and-oil with an explosive in suitable spots, and in some an explosive alone: and all I timed for ignition at midnight of the seventh day.

Hot now and sooty, I moved through the town, stopping regularly at every hundredth door: and I laid the faggots of conflagration, timing them for ignition at midnight of the seventh day.

* * *

Whatever door I found closed against me I drove at it with venom.

* * *

Shall I commit it to paper? that deep secret of the human organism? . . . As I worked, I waxed wicked as a demon! and with lowered neck, and outpush of the belly, and the blasphemous strut of tragic play-actors, I went: for here was no harmless fireworks, but the crime of arson, and a devilish, though vague, malevolence, and the rage to grind and raven and riot was upon me like a dog-madness, all the mood of Nero and Nebuchadnezzar, out of my mouth proceeding all the obscenities of the slum and gutter, and I sent up such hisses and giggles of challenge to Heaven that day as never yet has man let out. But this way lies frenzy . . .

* * *

I was angered, however, that day of the faggot-laying, even in the midst of my feeling of omnipotence, by the slowness of the motor, which made me kick it; and at that hill near the Old Dover Road the thing refused to move, the train too heavy for my horse-power: so there I stood impotent; no other motor visible, and most of the motors with exhausted accumulators, ruined magnetos, choked needle-valves, waterless or petrolless; there was a tram just there, but the notion of setting-up an electric station, with or without automatic stoking-gear, presented so hideous a picture of labor to me, that I would not entertain it. After half an hour, however, I remembered seeing hereabouts a power-station driven by turbines: so I uncoupled the motor, covered the drays with the tarpaulins, and went driving about, not caring whom I crushed; and, presently finding the station in a by-street, I went in by a window, a rage upon me to have my will quickly accomplished. I got some cloths and dusted a commutator; ran and turned the water into the turbines; set the lubricators running on the bearings; adjusted the generator-brushes; and ran up to the gallery to switch the current on to the line. By this time it was getting dark: so I hurried out, got into the car, and was off down three by-streets, till I turned into my own street; but had no sooner reached it than I pulled up with a jerk, with a shout of astonishment—the cursed street all lit up and gay! three arc-lamps not far apart revealing every feature of a field of dead; and there was a thing there the grinning impression of which I shall carry to my grave, a thing which spelled at me, and ceased, and began again, and ceased, and spelled at me: for above a shop was a flag, red with white letters, fluttering on the gale the name "Metcalfe's Stores;" and under the flag, stretched across the house, was the thing which spelled, letter by letter, in letters of brilliance deliberately, coming to an end, and going back to begin again, spelling

DRINK
ROBORAL

And that was the last word of civilized Man to me, Adam Jeffson—its ultimate gospel and message: *Drink Roboral!*

I was put into such a passion by this ribaldry, which affected me like the laughter of skeletons, that I rushed from the motor, threw two of my fuses at it, then went looking for stones to stone it; but no stones: and I had to stand there enduring that rape of my eyes, its dogged iteration, its taunting ogle—D,R,I,N,K R,O,B,O,R,A,L.

It was one of those advertisements worked by a little motor driven by the station, I had now set it going, and this nonsense stopped my operations for that day, since it was late: so I drove to the hotel which I had made my home in sullen and weary mood: for I knew that Roboral would not cure the least of all my sores.

* * *

The next morning I awoke in another frame of mind, disposed to idle, and let things slide. After washing in cold rose-water and descending to the *salle-à-manger,* which I had laid my morning-meal the previous evening, I promenaded the only one of these sombrous tufted corridors in which were not more than two dead, though behind the doors on both sides I knew that they lay in plenty. When I was warmed, I again went down, got four cans from other motors, and drove away—to Woolwich, as I thought: but instead of crossing the river by Blackfriars, I went more eastward, and, having passed into Cheapside, which was impassable, unless I crept, was going to turn back, when I observed a phonograph-shop, into which I got by a side-door, seized by curiosity to hear what I might hear: so I put one, with a lot of records, into the car, for there was still a strong peach-odor in this closed shop which

Listening to phonograph cylinders

displeased me; then proceeded westward through by-streets, seeking some house into which to go from the winds, when I saw the Parliament-house: and thither I went with my two parcels, walking into this old place along dusty busts, to deposit my boxes on a table beside a brass thing lying there, what they called "the Mace;" and I sat to hear.

Unfortunately, the phonograph was a clock-work one, and, when I wound it, would not go, so that I got angry, nearly tore it to pieces, and was half for kicking it; but there was a man seated in the chair which they called "the Speaker's Chair," who was in such a posture, that he had, every time I glanced at him, an air of bending forward with interest to watch what I was doing, a Mohrgrabim sort of man, almost black, with Jewish nose, crinkled hair, keffie, and flowing robe, probably a Galla, present with him being seven people only about the benches, mostly leaning forward with rested head, so that this room had quite a void and solitary mood; but this Galla, or Bedouin, with his grotesque interest in my doings, restrained my hands; and at last, by dint of poking and dusting, I got the phonograph to go.

And all that morning, and far into the afternoon, forgetful of food and of the cold which gradually possessed me, I sat there listening, musing—cylinder after cylinder: frivolous songs, orchestras, voices of famous men whom I had talked with, and shaken their solid hands, talking afresh to me, though rather thick-tongued and hoarse, from out of the vague void beyond the grave, most strange; and the fourth record that I put on, ah, I knew with a start that throat of thunder, knew it well: the "parson," Mackay's. . . . Over and over I heard that day those words of his, spoken when the cloud had got to the longitude of Prague: and in all that torrent of oratory not one note of "I told you so:" but he calls; ". . . praise Him, O Earth, for He is He; and if He slay me, I will laugh raillery at His sword, and banter Him to His face: for His sword is sharp Joy, and His poisons end my death. Care not, therefore, but take my comfort to your heart tonight, and my honeys to your tongue: for of old He chose thee, and once mixed spousally with thee in an ancient couch, O Afflicted: and He is thou, flesh of thy flesh. Hope, therefore, most at the nadir of despair: for He is nimble as a weasel, and He twists like quicksilver, and His tropics and turning-points are inborn in Being, and when He falls He falls like harlequin and shuttle-cocks, shivering plumb to His feet, and each third day, lo, He is risen again, and His defeats are the rough scaffolding from which He heaps His Parthenons, and the last end of this sphere shall be no peach-cloud, I say to you, but Carnival and Harvest-home. . . ."

* * *

So Mackay, with thick-tongued effort. I found this brown

room of the Commons-house, with its green benches, and grilled galleries, so agreeable to my mood, that I went again the next day, and listened to more records until they wearied me: for what I had was an itch to hear scandals and revelations of the festering heart, but these records, got from a shop, divulged nothing. I then went out to make for Woolwich, but, seeing in the motor the poet's note-book in which I had written, I took it, went back, and was writing an hour, till I was tired of that, too; and, judging it too late for Woolwich that day, wandered about the dusty committee-rooms and recesses of this considerable place. In one room another foolishness seized upon me, shewing how my whims have become more imperious within me than all the laws of the Medes: for in that Room No. 15 I found a young policeman, flat on his back, who pleased me: his helmet tilted under his head, and near one white-gloved hand an official envelope, that stagnant quiet apartment still peach-scented, and he gave not the slightest smell, though he was stoutish, his countenance now the color of ashes, in each cheek a hole large as a sixpence, his lids flimsy, vaulted, fallen into their caverns, from under whose fringe of lash was whispered the word "*Eternity.*" His hair seemed long for a policeman, probably had grown since death; but what interested me about him was the envelope at his hand: for "what," I asked myself, "was this fellow doing here with an envelope at three on a Sunday afternoon?" and this causing me to look closer, I saw by a mark at the left temple that he had been shot or felled; whereupon I was thrown into a rage, thinking that this poor man had been killed in the execution of his duty, when many had fled their post to pray or riot: so I said to him: "Well, D. 47, you sleep very well; and you did well, dying so; I am pleased with you, and decree that by my own hand you shall be distinguished with burial:" and this wind so possessed me, that I at once went out, and with the crowbar and spade from the car went into Westminster Abbey, where I routed up a grave-slab in the south transept, and began to dig; but, I do not know how, before I had digged a foot the impulse forsook me, so I left off, promising to resume it; but nothing was ever done, for the next day I was at Woolwich, and busy enough about other things.

* * *

During the next four days I worked with a fever on me, and a map of London before me.

There were places in that city!—secrets, vastnesses, horrors! In the wine-vaults at London Docks was a vat which must have contained twenty thousand gallons, and with a dancing heart I laid a train there; the tobacco-warehouse must have covered eighty acres, and there I laid a fuse; in a house near Regent's Park, shut in from the street by shrubbery and a wall, I saw a thing . . . ! and what shapes a great city hid I only now know.

* * *

I left no quarter unremembered, taking a train of eight vehicles, now drawn by three motors, with which I visited West Ham and Kew, Finchley and Clapham, Dalston and Marylebone; deposited piles in the Guildhall, in Holloway Gaol, in the Tower, in the Parliament-house, in St. Giles' Workhouse, under the organ of St. Paul's in the Kensington Museum, in Whiteley's place, in the Trinity House, in the Office of Works, in the recesses of the British Museum; in a hundred warehouses, in five hundred shops, in a thousand dwellings. And I timed them all for ignition at midnight of the 23rd of April.

By five in the afternoon of the 22nd, when I left my train in Maida Vale, and drove alone to the house on high ground near Hampstead Heath which I had chosen, the thing was finished.

* * *

The morning dawned, and I was early astir: for I had much to do.

I intended to make for the coast the next day, so had to select a motor, store it, have it in a place of safety; and I had to tow another vehicle stored with trunks of fuses, books, clothes, and other little things.

My first journey was to Woolwich, whence I took all that I might ever want in the way of mechanism; thence to the National Gallery, where I cut from their frames the "Vision of St. Helena," Murillo's "Boy Drinking," and "Christ at the Column;" and thence to the Embassy to bathe, anoint my body, and dress.

As I had anticipated and hoped, a blustering storm was blowing from the north.

When I had started out from Hampstead at 9 a.m. I had had to assume that some of my fuses had somehow forerun, for I saw morose hazes at various points, was anon aware of the dumb bump and hum of some explosion somewhere remote, as when gunnings of Mont Pelée in Martinique get, like echoes of God's voice, to ears in Guadeloupe; and by noon I felt sure that several regions of east London must already be in flames. With the solemn feelings of bridegrooms and marriage-mornings—with a shrinking heart, God knows, yet a heart upbuoyed on thrilling joys—I drove about on the business of the orgy of the night.

* * *

The house at Hampstead, which no doubt still stands, is of agreeable design in quite a stone and rural style, with breadths of wall-surface, two coped gables, mullioned windows, oversailing verge-roofs; but, rather spoiling it, a tower at the south-east corner, on the top floor of which I had slept the previous night. There I had a box of pallid tobacco compounded with rose-leaves and opium, found in a foreign house in Seymour Street, also a true Saloniki hookah, with Cyclades wine, nuts, and so on, and a gold harp stamped with the name of Krasinski, taken from his house in Portland Street.

But so much did I find to do that day, so many odd things turned up which I thought that I would take, that it was not until six that I drove finally northward through Camden Town. And now an awe possessed my soul at the solemn noise which everywhere encompassed me, an ineffable awe, a blessed terror. Never could I have dreamt of aught so great and strong. Everywhere over my head there rushed southward with outstretched throat and a wing of haste a smoke inflamed; and, mixed with the roaring, I heard hubbubs of tumblings and rumblings, unaccountable, like the moving-about of furniture in the houses of Titans, while pervading the air was a most weird and tearful crying, as it were threnody and nenia, and wild wails of pain, dying swan-songs, and all lamentations at cosmic break-up and tribulation. Yet I was aware that at such an hour the flames must be far from general; in fact, they had not well commenced.

* * *

As I had left a region of houses without combustibles to the south of the house which I was to occupy, and as the storm was from the north, I simply left my two vehicles at the door without fear; then went up the tower, lit the candles, ate voraciously of the dinner which I had left ready; then with hands which shivered arranged the bed-clothes upon which to drop in the morning hours, opposite the wall where the bed was being a Gothic window, large, with a low sill, looking south: so that I could recline at ease in the easy-chair, and see. It had been a young lady's room, for on the toilette were crystals of Laliqué, a plait of brown hair, powders, *rouge-aux-lèvres,* one little bronze slipper, and knickknacks, and I loved her and hated her, though I did not see her anywhere; anyway, before nine I was seated at the window to watch, all being ready at my hand, the candles extinguished: for the theatre was opened, and the atmosphere of this earth seemed turned into Hell, and Hell was in my soul.

* * *

Immediately after midnight there was a visible increase in the conflagration, when on all hands I began to see structures soar ablaze, with grand hurrahs, on high, in fives and tens, in twenties and forties: all between me and the limit of my vision they leapt, then lingered, they fell, while my spirit more and more felt—deeper mysteries of sensation, sweeter thrills. I sipped exquisitely, I drew out enjoyment leisurely. Anon when some more expansive angel of flame arose with steady aspiration, to tarry with spread arms, and scatter, I would lift a little to clap, as at acting, or would call to them in the names of woman with "higher, wild Polly," "hop, Cissy, you flea," or "Bertha, burst:" for now I seemed to see pandemonium through crimson spectacles, the air wildly hot, and my eyeballs like theirs that walk staring in the midst of burning fiery furnaces, and my skin itched with a rich and prickly itch. Anon I touched the chords of the harp to the air of Wagner's "Walkürenritt."

Near three in the morning I reached the climax of my wicked sweets, my drunken eyelids closing in a luxury of pleasure, and my lips lay stretched in a smile that drivelled; a feeling of dear peace, of power without bound, consoled me: for now the whole field at which through streaming tears I peered, mustering its hundred thousand thunders, and brawling beyond the clouds the voice of its southward-bounding torment, wobbled to the horizon one ocean of smokeless fire, in which sported and washed themselves all that dwell in Hell, with callings, flights, and holiday; and I—first of my kind—had flashed a sign to the nigher planets. . . .

* * *
* * *

Those words "nigher planets" I wrote thirteen months ago, some days after the destruction of London, I being then on board the old *Boreal,* bound for France; for the night was dark, though calm, and I was frightened of running into some ship, so wrote to do something, the ship lying still; and though the book in which I wrote has been with me, no impulse to scribble has since visited me, until now.

I had no intention of wearing out my life in lighting fires in that island, and came to France with the idea of seeking some palace in the Riviera, Spain, there for the present to make my home: so I set out from Calais toward the end of April, taking my things, by train at first, then, being in no hurry, by motor, maintaining a south direction, ever anew astonished at the luxuriance of the forest vegetation which within so short a space chokes this pleasant land, even before the advent of summer.

After three weeks of slow travelling—for France with her pavered villages, hilly character, forests, and country-manner, is always charming to me—after three weeks I happened upon a valley which had never entered my head, and the moment I saw it I said: "Here I will live," though I had no idea what it was, for the monastery which I saw did not look like a monastery, according to my ideas: but the map showed that it must be La Chartreuse de Vauclaire in Périgord.

In France, at Vauclaire -

This word "Vauclaire" must be a corruption of *Vallis Clara*, for *l*'s and *u*'s did interchange about in this way—"fool" and "fou"—which proves the dear laziness of French people, for the "l" was too much trouble for them to sing: at any rate, this Vauclaire, or Valclear, was well named, for here, if anywhere, is Paradise, and, if anyone knew how and where to build and brew liqueurs, it was those old monks, who followed their Master with *entrain* in that Cana miracle, but æsthetically shirked to say to any mountain: "Be thou removed."

* * *

The hue of the vale is cerulean, resembling that blue of the robes of Albertinelli's Madonnas, the monastery itself consisting of an oblong space, or garth, round three sides of which stand sixteen small houses, all identical, cells of the fathers, looking inwards upon cloisters; and in one part of the oblong, under cypress sighings, black crosses over graves.

West is the church, the hostelry, a court with some trees and a fountain; and, beyond, the entrance-gate.

All this on a slope green as grass, backed against a mountain-side of which one does not see the tree-trunks, the trees resembling one leafy tree-top, run out over the breadth of the mountain's breast.

* * *

I was there four months, till something drove me away. What had become of the brothers I do not know, for I only found five, four of whom I took in two journeys in the motor to the church of Saint Martial d'Artenset, and left them there; and the fifth remained three weeks with me, for I would not remove him from his prayer: a brother who knelt in his cell robed and hooded in his phantom white, for like whatever is most phantom, visionary, eerie must a procession of these people have seemed at evening or midnight; he in his pigmy chaste chamber glaring upward at his Christ, who hung long-armed in a recess beside three book-shelves; under the Christ a Madonna, gilt-and-blue; the books on the three shelves few, leaning different ways; his elbow on a table at which was a chair; and, behind him, in a corner, the bed—a bed all en-nooked in boards: two perpendicular boards at foot and head, reaching the ceiling, a horizontal board at the side over which he got into bed, and another like it above it for fringe, making the bed within a shady den. He was a big severe being, forty, blond as corn, but with red also in his hairy beard; and appalling was the significance of that glare that prayed, and the long-drawn gauntness of those jaundiced jaws. I cannot explain to myself my reverence for this man; but I had it.

It was my way to plant at the portal the carved chair from the chancel on sultry days, and rest my soul, refusing to meditate on any thing, drowsing and smoking for hours: all down there in the plain being woods of fruit waving about the prolonged thread of the river Isle, whose route winds loitering quite nigh the foot of the monastery-slope; this slope dominating the village of Monpont all in thicket, the Isle drawing its waters through the village-meadow, which is dim with shades of oaks: and to have played there a boy, using it familiarly as one's own breathing and foot, must have been pretty sweet and homely.

Well, one morning after four months I opened my eyes in my cell to the piercing consciousness that I had burned Monpont over-night: and so overcome was I with compunction for this poor inoffensive little place, that for two days, scarcely eating, I paced between the oak pews of the nave—massive stalls they are, separated by Corinthian pilasters—wondering what was to become of me, and if I was not already mad; and there are some little angels with strangely human faces, Greuze-like, supporting the nerves of the apse, which, every time I passed them, seemed conscious of me and my existence there; and the woodwork which ornaments the length of the nave, and of the choir also, all an intricacy of marguerites and roses, here and there took in my eyes significant forms from particular points of view; and there is a partition—for the nave is divided into two chapels, one for the brothers and one for the fathers—and in this partition a massive door, which yet looks quite light and graceful, carved with oak and acanthus leaves, and every time I passed through I had the impression that the door was a sentient being, sub-conscious of me; and the Italian-Renaissance vault which springs from the nave seemed to look upon me with a gloomy knowledge of me, and of the heart within me: so that in the afternoon of the second day, after pacing the nave for hours, I dropped down at one of the two altars near that door of the screen, entreating God to have pity upon my soul; and in the very midst of my praying, I was up and away, the devil in me; leapt into the motor; nor did I come back to Vauclaire for another month, and came leaving regions of desolation in my rear, cities, furnaces of timber, Bordeaux burned, Lebourne burned, Bergerac burned.

* * *

I returned to Vauclaire, for it seemed now my home; and there I experienced a true, a deep repentance; and I humbled myself before my Maker. In which state I was seated one day in front of the monastery-gate when something said to me: "You will never be a good man, nor permanently escape Hell and frenzy, unless you have an aim in life, devoting yourself heart and soul to some work which will exact all your thought, your ingenuity, your knowledge, your strength of body and will, your skill of head and hand: otherwise you are bound to succumb. Do this, therefore, beginning, not tomorrow, but now: for though no man will see your work, there is still the

Almighty God, who is also something in His way: and He will see how you strive, and try, and groan; and perhaps, seeing, may have mercy upon you."

* * *

In this way arose the notion of the palace—a notion, indeed, which had entered my brain before, but merely as a bombastic outcome of my mad moods: now, however, in a very different way, soberly, and, before long, occupying itself with details, difficulties, means, limitations, and every species of practical matter-of-fact; and every obstacle which, one by one, I foresaw was, one by one, as the days passed, overborne by the ardor with which that notion, soon becoming a mania, possessed me. After nine days of incessant meditation, I decided Yes; and I said: I will build a palace which shall be both a palace and a temple: the first human temple somewhat worthy the Potency of Heaven, the only human palace worthy the satrap of earth.

* * *

After this decision I remained at Vauclaire another week, a different man from the lounger it had seen, strenuous, converted, humble, making plans of this and of that, of the detail, and of the whole, drawing, multiplying, adding, conics, fluxions, graphs, totting up the period of building, which came out at a little over twelve years, estimating quantities and strength of material, weight and bulk, my nights full of nightmare as to the *kind*, deciding as to the size and structure of the crane, forge, and workshop, and the necessarily-limited weights of their parts, making a catalogue of over 2,400 articles, and finally, up to the fourth week after my departure from Vauclaire, skimming through the topography of almost the whole globe, before fixing upon the island of Imbros for my site.

* * *

I went back to England, and once again to those vacant windows and heaped black streets of what had been London: for its bank-vaults, &c., contained the necessary complement of the gold brought by me from Paris and then stowed in the *Speranza* at Dover, nor had I sufficient familiarity with French industries and methods to ferret out, even with the help of *Bottins,* one half of the 4,000 odd objects which I had now catalogued. My ship was the *Speranza,* which had brought me from Havre: for at Calais, to which I first journeyed, I could discover nothing suitable for all purposes, the *Speranza* being an American yacht, palatially fitted, three-mastēd, air-driven, with a carrying capacity of 2,000 tons, Tobin-bronzed, in good condition, containing sixteen interacting tanks, with a six-block pulley-system amidships which enables me to lift considerable weights without the aid of the hoisting air-engine, high in the water, sharp, handsome, having-in a few tons only of sand-ballast, and needing when I found her only three days' swotting at the water-line and engines to make her decent and fit: so I tossed out her dead, backed her from the Outer to the Inner Basin to my train on the quay, took in the sixty-three hundred-weight-bags of gold, and the half-ton of amber, and with this alone went to Dover, to Canterbury by motor, and thence by train, with a store of explosives for blasting obstructions, to London, proposing to make Dover my dépôt, and the London rails my thoroughfare from every direction of the country.

But instead of four months, as I had estimated, it kept me ten, a harrowing slavery: I had to blast no less than twenty-five trains from the path of my laden wagons, several times blasting away the metals as well, and then having to travel hundreds of yards without metals: for the labor of kindling the obstructing engines, to shunt them down sidings perhaps distant, was a thing which I would not undertake. However, all's well that ends well, though, if I had it to go through again, no, I should not. The *Speranza* is now lying nine miles off Cape Roca, a mist on the still sea, this being the 19th of June at ten in the night; no wind, no moon; cabin full of mist; and I pretty listless and disappointed, wondering in my soul why I was such a fool as to wallow in all those toils, ten long servile months, my good God, and now gravely thinking of throwing the whole thing to the devil; she pretty deep in the sea, pregnant with the palace. When the thirty-three . . .

* * *
* * *
* * *
* * *
* * *

Those words "when the thirty-three" were written over seventeen years since—long years—seventeen in number, nor have I now any notion to what they refer. The book in which I wrote I had lost in the *Speranza* cabin, and yesterday, in coming home to Imbros from an hour's cruise, found it there behind a chest.

I now find considerable difficulty in guiding the pencil, and these lines now written have quite an odd look, like the handwriting of a man not proficient in the art: it is seventeen years . . . Nor is the expression of my ideas fluent—have to think for the word, and I should not be surprised if the spelling is queer: I have been thinking inarticulately perhaps all these years; and now the letters have rather a foreign air to me, like Russian; or perhaps it is my fancy: for that I have fancies I know.

But what to write? The history of those seventeen years

could not be put down, my good God; at least, it would take seventeen more to do it. If I were to detail the building of the palace alone, and how it killed me nearly, and how I twice fled from it, and had to come back, and became its bounden slave, and dreamt of it, and grovelled before it, and prayed, and raved, and rolled; and how I forgot to make provision in the north wall for the expansion of the gold in summer, and had to break down eight months' work, and how I cursed Thee, how I cursed Thee; and how the lake of wine evaporated faster than the conduits replenished it, and the five voyages which I had to take to Constantinople for loads of wine, and my frothing despairs, till I had the thought of placing the reservoir in the platform; and how I had then to break down the south side of the platform to the very bottom, and the prolonged nightmare of terror that I had lest the south side of the palace would undergo subsidence; and how the petrol failed, and of the three-weeks' search for petrol along the coast; and how, after list-rubbing all the jet, I found that I had forgotten the rouge for polishing; and how, in the third year, I found the fluate for water-proofing the pores of the platform-stone nearly all leaked away in the *Speranza's* hold, and I had to get silicate of soda at Gallipoli; and how, after two years' observation, I had to come to the conclusion that the lake was leaking, and discovered that this Imbros sand was not suitable for mixing with the skin of Portland cement which covered the cement-concrete, and had to substitute sheet-bitumen in three places; and how I did all, all, for the sake of God, thinking "I will work, and be a good man, and cast Hell from me; and when I see it stand finished it will be an Altar and a Testimony to me, and I shall find peace, and be well:" and how I have been cheated—seventeen years, long years of my life—for there is no "God;" and how my plasterers'-hair failed me, and I had to use flock, hessian, scrym, wadding, whatever I could find, for filling the spaces between the platform cross-walls; and of the espagnolette bolts, how a number of them strangely vanished, as if snatched to Hell by harpies, and I had to make them; and how the crane-chain would not reach two of the silver-panel castings when finished, and they were too heavy for me to lift, and the wringing of the hands of my despair, and my dragging up of the grass, and the transport of my wrath; and how, for all one wild fortnight, I sought in vain for the text-book which describes the ambering process; and how, when all was all but over, in the blasting away of the forge and crane with gun-cotton, a crack appeared down the gold of the east platform-steps, and how I would not be consoled, but mourned and mourned; and how, in spite of all my sorrows, it was divine to watch my power grow from its troglodyte-beginnings of hundredweights, until I could swing tons, squeeze the flowing metals between the mould-end levers and the plungers, build at ease in a travelling-cage, and through sleepless hours view from my hut-door under the moon's electric-light of this land the three piles, of gold stones, of silver panels, of squares of jet, and be comforted; and how the putty-wash—but is is over: and not to live over again that vulgar nightmare of means and ends have I taken to this writing again—but to write down something, if I dare.

Seventeen years, my good God, of that delusion! I could put down no sort of explanation for all those groans and griefs at which a reasoning being would not shriek with derision, for I should have lived at ease in some retreat of the Middle-Orient, and burned my cities: but no, I must be "a good man" —vain notion. The words of a turbulent madman, that "parson" man in Britain who predicted what happened, were with me, where he says "the defeat of Man is *His* defeat;" and I said to myself "Well, the last man shall not be quite a fiend, just to spite That Other;" and I worked and groaned, saying "I will be a good man, and burn nothing, nor utter aught unseemly, nor debauch myself, but choke back the blasphemies that Those Others shriek through my throat, and build and build, with griefs and groans;" and it was vanity: though I do love the house, too, I love it well, for it is my home in the waste.

I had calculated to finish it in twelve years, and I should have finished it in fourteen, but one day, when the south and west platform-steps were already finished—it was in the July of the third year, near sunset—as I left off work, instead of stepping to the tent where my dinner lay ready, I paced down to the ship—strangely—in a daft, mechanical kind of way, without saying a word to myself, a smile of malice on my lips; and at midnight was lying off Mitylene, thirty miles south, having bid, as I thought, a last goodbye to all those toils. I was going to burn Athens.

I did not, however; but kept on my way westward round Cape Matapan, intending to destroy the forests and towns of Sicily, if I found there a suitable motor for travelling, for I had not been at the pains to take the motor on board at Imbros; otherwise I would ravage parts of southern Italy. But when I came thereabouts I was confronted with a horror: for no southern Italy was there, and no Sicily was there, unless a little island five miles long was Sicily, for nothing else I saw, save the crater of Stromboli, smoking still; and, as I cruised northward, looking for land, for a long time I would not credit the evidence of the instruments, thinking that they wilfully misled me, or I stark mad. But no: no Italy was there, until I came to the latitude of Naples, it too, having vanished, engulfed, engulfed, all that stretch; from which monstrous thing I got so solemn a shock and mood of awe, that the mischievous mind in me was quite chilled and quelled, for it was, and is, my belief that widespread re-arrangement of the earth's surface is being purposed, and in all that drama, O my God, how shall I be found?

However, I went on my way, but more leisurely, not daring during many days to do anything, lest I might offend anyone; and, in this foolish cowering mood, coasted all the west coast of Spain and France during seven weeks, in that prolonged intensity of calm which at present alternates with storms that transcend all thought, till I came again to Calais: and there, for the first time, landed.

Here I would no more contain myself, but burned; and that stretch of forest between Agincourt and Abbéville, five square miles, I burned; and Abbéville I burned; and Amiens I burned; and three forests between Amiens and Paris I burned; and Paris I burned; burning and burning during four months, leaving in my rear reeking regions, a tract of ravage, like some being of the Pit that blights where his wings of fire pass.

* * *

This of city-burning has now become a habit with me more enchanting—and more debased—than ever was opium to the smoker; my necessary, my brandy, my bacchanal, my secret sin. I have burned Calcutta, Pekin, and San Francisco. . . . In spite of the curbing influence of this building, I have burned and burned—three hundred cities and countrysides. Like Leviathan disporting himself in the sea, so I have rioted in the earth.

* * *

After an absence of six months, I came back to Imbros: for I was for gazing again upon the building that I had done, that I might mock myself for all that unkingly grovelling; but when I saw it, standing there as I had abandoned it, frustrate and forlorn, waiting its maker's hand, some pity and instinct to build took me: for something of God was in man; and I dropped prostrate, and spread my arms to God, and was converted, promising to finish the work, with prayers that as I built so He would build my will, and save the last man from the enemy. And I set to work that day to list-rub the last six dalles of the jet.

* * *

I did not leave Imbros after that during four years, except for brief trips to the coast—to Kilid-Bahr, Gallipoli, Lapsaki, Gamos, Erdek, Erekli, once even to Constantinople—if I happened to want anything, or was weary of work, but without once doing the least hurt to anything, containing my humors, and fearing my Maker; and full of peace and charm were those cruises through this Levantic world, which, truly, is rather like a sketch in water-colors done by an angel than like the dun earth; and full of self-satisfaction and pious contentment would

I cruise back to Imbros, approved by my conscience, for that I had evaded temptation, and lived tame and stainless.

I had set up the southern of the two closed-lotus columns, and the platform-top was already looking as lovely as heaven, flushing its glory of two-foot squares, pellucid jet alternating with pellucid gold, when I noticed one morning that the *Speranza's* bottom was really now too foul, and the caprice seized me then and there to leave everything, and clean her: so I went on board, descended to the hold, threw off my sudeyrie, and began to shift the ballast over to starboard, to tilt up her port bottom to the scraper: wearying labor, and about noon I was seated on a ballast-bag, resting in the semi-gloom down there, when something seemed to whisper into me: *You dreamed last night that there is an old Chinaman alive in Pekin.*" Horridly I started: I *had* dreamed something of the sort; and I sprang to my feet.

I cleaned no *Speranza* that day; nor for three days did I anything, but sat on the cabin-house brooding, my palm among the hairy draperies of my chin upholding it: for the notion of such a thing, if it could by any possibility be true, was detestable as death to me, changing the color of the sun and the whole tone of existence; and anon at the outrage of that thing my brow would flush with rage, and my eyes blaze; till in the fourth twilight I said to myself: "That old Chinaman in Pekin is likely to be devoured by fire, I think, or be blown to the clouds."

So, a second time, on the 4th of March, the poor palace was left to build itself: for, after a trip ot Gallipoli, where I got some lime-twigs in boxes of earth, and some preserved limes and ginger, I set out for a long voyage to the East, passing through the Suez Canal, and visiting Bombay, where I was three weeks, and then destroyed it.

* * *

I had the thought of travelling across Hindostan by engine, but did not wish to leave my ship, to which I was attached, not sure of seeing anything so suitable at Calcutta; and, moreover, I was afraid to abandon my motor, which I had taken on board with the air-windlass: I therefore went down the west coast.

All that northern shore of the Arabian Sea has at present an odor which it wafts far over the ocean, resembling odors of happy dream-lands, sweet to smell in the early mornings as if the earth were all a perfume, and heaven and inhalation.

On that voyage, however, I had, from beginning to end, twenty-seven fearful storms, or, if I reckon that one near the Carolines, twenty-eight; but I do not wish to write of these rages: they were too inhuman; and how I came alive through them against my wildest hope Someone, or Something, only knows.

I will put down here a thing; it is this, my God—something that I have noted: a definite obstreperousness in the temperament of the elements now, when once roused, which grows, which grows. Tempests have become very far more wrathful, the sea more truculent and unbounded in its insolence; when it thunders it thunders with a rancor new to me, cracking as though it would wreck the welkin's vault, and bawling through the heaven of heavens as if roaring to devour all being; in Bombay once, in China thrice, I was shaken by earthquakes, the second and third marked by a certain extravagance of agitation that might turn a man grey. Why should this be, my God? I remember being told ages ago that on the American prairies, which of old had been swept by great tempests, the tempests gradually subsided when man went to reside there: so, if this be true, it would seem that the mere presence of man had a certain subduing or mesmerising effect on the innate turbulence of Nature, and his absence today may have taken off the curb. It is my belief that within fifty years the forces of the earth will be turned fully loose to tumble as they choose, and this globe will become one of the undisputed playgrounds of Hell, the theatre of commotions huge as those witnessed on Jupiter.

* * *

The earth is all on my brain, on my brain, O dark-minded Mother, with thy mighty cravings, thy regrets, and bleak griefs, and comatose slumbers, and doom to come, O Mother, and I, poor man, though a monarch, the one witness of the drama of thy monstrous sorrows. Upon her I brood, and do not stop, but brood and brood—the habit, I think, becoming fixed and fated during that Orient cruising: for what is in store for her God only knows, and I have seen in my broodings visions of her future, which, if a man should see with the instrument of flesh, he would spread the arms, and wheel and wheel through the mazes of a giggling frenzy, for the vision by itself is the brink of giddiness. If I might stop but for one hour from brooding upon her! but I am her child, and my mind grows and grows to her like the off-shoots of the banyan-tree, which strike back downward to take root, and she sucks and draws it, as her gravity draws my foot, and I cannot take flight from her, for she is greater than I, and there is no escaping her: so that in the end, I know, my soul will hurl itself to ruin, like erring sea-birds upon pharos-lights, against her wild and mighty bosom. A whole night through I may lie open-eyed, my brain obsessed with that Gulf of Mexico, how identical its hollow with the protuberance of Africa opposite, and how the protuberance of Brazil fits-in with the hollow of Africa: so that it is obvious to me—*obvious*—that they once were one, and one night shied so far apart; of which thing the wild Atlantic was aware, and ran blithely, hasting in between: and how if an eye had been there to see, and an ear to hear that solitary oratory of Thee . . . Thou, Thou . . . and if now again they throw-together, so long divorced . . . but that way fury lies. Yet one cannot but think: for she fills my soul, and absorbs it, with all her moods and ways. She has meanings, secrets, plans. . . . Strange that twinness between the scheme of Europe and the scheme of Asia: each with three southern peninsulas, three twins; Spain, Arabia, Italy-Sicily, India-Ceylon; the Morea and Greece split by the Gulf of Corinth, the Malay Peninsula and Annam split by the Gulf of Siam; each with two northern peninsulas pointing south—Sweden and Norway, Korea and Kamschatka; each with two island-twins—Britain, Japan; the Old World and the New has each a peninsula pointing north—Denmark, Yucatan: Denmark a forefinger with long nail, Yucatan a thumb—pointing to the Pole. What does she mean? Is she herself a living entity with a will and a fate, as sailors said that ships were? And that thing that wheeled at the Arctic, wheels it still away there in its dark ardour? Queer that volcanoes are all near the sea: I don't know why. This fact, added to the fact of submarine explosions, used to support the chemical theory of volcanoes, which supposed the infiltration of the sea into ravines containing the materials which make the fuel of eruptions: but God knows if that is true. The lofty ones are intermittent—a century, three, ten, of still waiting, and then their oratory struck dumb for ever some poor district; the low ones are constant; and sometimes they form a linear system, consisting of vents, like the chimneys of some foundry underneath. Who could know the way of her? In mountains, a series of peaks denotes the presence of dolomites; rounded heads mean calcareous rocks; needles mean crystalline schists: but why? I have some knowledge of her for ten miles down, but whether through eight thousand miles she is flame or small-shot, hard or soft, I do not know, I do not know. Her method of forming coal, geysers, and sulphur-springs, and the jewels, the atols and coral reefs; the rocks of sedimentary origin, like gneiss; the plutonic rocks, rocks of fusion, and the unstratified rubble that constitutes the basis of the crust; and harvests, the flame of flowers, and the passage from the vegetable to the animal: I do not know them, but they are of her, and are like me, molten in the same oven of her scarlet heart. She is dark and moody, sudden and ill-fated, and tears her young like a cannibal cat; and she is old and deep, and remembers Ur of the Chaldees that Uruk erected, and the amœba's first stir, and remembers that Temple of Bel, and bears still as a thing of yesterday old Persepolis and the tomb of Cyrus, the site of Haran, and those vihârah-temples hewn from the Himalayan stone; and, in coming home from the Orient, I stopped at Ismailia, and so to Cairo, saw where Memphis was, and brooded one midnight before that pyramid and that mute sphynx, seated in a tomb, until tears of pity streamed down my cheeks: for man "passeth away." These rock-tombs have columns extremely like the two palace-pillars, only that these are round, and mine are square, but the same band near the top, over this the closed lotus-flower, then the plinth that separates them from the architrave, only mine have no architrave, the tombs themselves consisting of an outer court, then comes a well, and inside another chamber for the dead; and there, till the want of food drove me away, I remained: for more and more the earth overgrows me, woos me, assimilates me; so that I ask myself this question: "Must I not in some years cease to be a man, to become a small earth, her copy, weird and fierce, half demoniac, half ferine, wholly mystic—morose and turbulent—fitful, and mad, and sad—like her?"

* * *

A month of that voyage, from May the 15th to June the 12th, I squandered at the Andaman Islands near Malay: for that any old Chinaman should be alive in Pekin commenced to appear the queerest whimsey that ever entered a head; and those jungled islands of the sun, to which I had got after a vast orgy one night at Calcutta, when I fired not merely the city but the river, pleased my fancy to such an extent, that at one time I meant to abide there, I being at the one named "Saddle Hill": and seldom have I had such sensations of peace as I lay a whole day in a valley, deep in the shadow of tropical ranknesses, watching the *Speranza* at anchor: for the valley rose from a bay, of which I could see one peak lined with cocoanut-trees, all cloud scorched out of the sky except the flimsiest lawn-figments, and the sea as still as a lake breathed on by breezes, yet making a considerable noise in its breaking on the coast, as I have noticed in these sorts of places: I do not know why. These Andaman people seem to have been quite savage, for I met some in roaming the island, nearly skeletons, yet with limbs still cohering, and in some cases mummified relics of flesh, and never a shred of clothes: a strange thing, considering their closeness to old civilizations: they looking small and black, or almost, and I never found a man without seeing near him a spear, so that they were keen folk, the earth's perversity spurred in them, too, and I was so pleased with these people, that I took on board with the gig one of their little tree-canoes: which was my foolishness: for gig and canoe were three days later smitten from the decks into the middle of the sea.

* * *

I passed down the Straits of Malacca, and in that short distance between the Andaman Islands and the S.W. corner of Borneo I was thrice so mauled, that at times it seemed out of the question that anything built by man could outlive such cataclysms; and I abandoned myself, but with bitter reproaches, to perish darkly, the effect of the last on me, when it was over, being the unloosening anew of my tumid moods: for I said "Since they mean to slay me, death shall find me rebellious"; and for weeks I did not sight some specially blessed village, or umbrageous spread of timber, that I did not stop the ship, and land the materials for their destruction: so that nearly all those odorous lands about the north of Australia will bear the traces of my hand for many a year: for more and more my voyage grew loitering and zigzagged, as some whim shunted it, or a movement of the pointer on the chart; and I thought of chewing the lotus of sloth and nepenthe, enchanted in some pensive nook of this summer, where from my hut-door I should see through the opal hues of opium the sea-lagoon gush sluggishly upon the coral atol, and the cocoanut-tree would droop like slumber, and the breadfruit-tree would mumble in dream, and I should watch the *Speranza* at anchor in the pallid atol-lake, year after year, and wonder what she was, and whence, and wherefore she dozed so deep for ever; and after an age of melancholy peace I should note that sun and moon had ceased to move, and hung spent, opening anon an eyelid to doze again, and God would sigh "Enough," and nod: for that any old Chinaman should be alive in Pekin was a thing so fantastically maniac, as to cast me at times into paroxysms of wild red laughter that left me faint.

During four months, from June into October, I visited the Fijis, where I saw heads still englobed in thickets of stiff hair; in Samoa skulls coroneted wtih nautilus-shell, and in one townlet an assemblage of bodies suggesting some festival: so that I believe that these people perished on a day of woe and overthrow without the least presage of anything. The women of the Maoris wore an abundance of jade embellishments, and I found a peculiar kind of shell-trumpet, one of which I have now, with a tattooing chisel and a wooden bowl nicely carved; while the New Caledonians went nude, confining their attention to the hair, wearing apparently an artificial hair made of the fur of some animal like a bat, and they wore wooden masks, and big rings—for the ear, no doubt—which must have reached down to the shoulders: for the earth urged them every one, and made them wild, wayward and various like herself. I went from one to the other without any system whatever, seeking the ideal resting-place, and frequently thinking that I had found it, only to weary of it at the feeling that there might be a yet deeper and dreamier in-being; but in this seeking I received a check, my God, which chilled me to the liver, and set me fleeing from these places.

* * *

One night, the 29th of November, I dined late—at eight—sitting, as was my way in calm weather, cross-legged on the cabin-rug in the starboard aft corner, a semicircle of *Speranza* gold-plate before me, and near above me the lamp's red glow and green conical reservoir, whose creakings never cease in the stillest mid-sea; and beyond the plates the array of soups, meat-extracts, meats, fruit, sweets, wines, nuts, liqueurs, coffee on the silver spirit-tripod—all which it was always my care to select from the store-room and lay out once for all in the morning. I was late, seven being my hour, for on that day I had been engaged in the job, always postponed, of overhauling the ship, brushing here a rope with tar, there a board with paint, there a crank with oil, rubbing a door-handle, a brass-fitting, filling the three cabin-lamps, dusting mirrors, dashing the plains of deck with bucketfulls, and, up aloft, chopping loose with its rigging the mizzen topmast, which for a month had been sprained at the clamps: all this in cotton drawers under my loose *quamis*, bare-footed, my beard knotted up, the sun ablaze, the sea smooth and pallid with that smooth pallor of currents in a hurry, the ship pretty still, no land near, yet large tracts of sea-weed reaching away eastward—I at it from 11 until near 7, when sudden darkness interrupted: for I wanted to have it all over in one offensive day: so I was pretty weary when I went down, lit the central lever-lamp and my own two, dressed in my room, then to dinner in the saloon; and voraciously I ate, perspiration, as usual, pouring down my brow, using knife or spoon in the right hand, but never the Western fork, licking the plates clean in the Mohammedan manner, drinking pretty freely. Still I was weary and went on deck, where I had the easy-chair with the broken arm, its blue-velvet threadbare now, before the wheel; and in it I lay, smoking cigar after cigar from the Indian D box, half-asleep, yet conscious, while the moon moved up into a sky nearly cloudless: and she was bright, but not bright enough to outshine that enlightened flight of the ocean, which that night was one swamp of phosphorescence, a wild luminosity of jack-o'-lantern thronged with stars and flashes—the whole trooping unanimously, as if in haste with some momentous purpose, an interminable assemblage teeming, careering eastward in the sweep of an urgent current. I could hear it in my sluggard slumbrousness struggling at the bound rudder, gulping sloppy noises of hogs' chops guttling beneath the sheer of the poop; and I knew that the ship was slipping along pretty quickly, drawn into the trend of that procession, probably at the rate of six knots; but I did not care, knowing very well that no land was within two hundred miles of my bows, for I was in long. 173°, in the latitude of Fiji and the Society Islands, between those two; and after a time the cigar drooped and dropped from my mouth, drowsiness overcame me, and I slept there, in the lap of immensity.

* * *

So that something preserves me: Something, Someone: *and*

Comes into vicinity of source of purple cloud

for what? . . . If I had slept in the cabin, I must most certainly have perished: for, stretched there on the chair, I dreamed a dream which once I had dreamed in snows yonder in the beyond of that hyperborean North: that I was in an Arab paradise; and I had a protracted vision of it, for I reached up amid the trees, and picked the peaches, and pressed the blossoms to my nostrils with breathless inhalations of fondness: until a sickness woke me, and when I opened my eyes the night was gloomy, the moon down, everything drenched with dew, the sky a jungle lush with stars, bazaar of maharajahs tiaraed, begums arrayed in garish trains, and all the air informed with that mortal afflatus; and high and wide uplifted before my sight—stretching from the northern to the southern limit—a row of eight or nine smokes, inflamed as from the chimneys of some Cyclopean forge which goes all night, most solemn, most great and dreadful in the solemn night: eight or nine, I should say, or it may be seven, or it may be ten, for I did not reckon them; and from those craters puffed up gusts of encrimsoned stuff, there a gust and there a gust, with tinselled fumes that convolved upon themselves, glittering with troops of sparks and flashes, all in a garish haze of glare: for the foundry was going, though languidly; and upon a land of rock four knots ahead, which no chart had ever marked, the *Speranza* drove straight with the sweep of the phosphorus sea.

As I rose, I fell flat: and what I did thereafter I did in a state of existence whose acts, to the waking intellect, seem unreal as dream. I must immediately, I think, have been conscious that here was the cause of the destruction of organisms, conscious that it still surrounded its own neighborhood with baneful emanations, conscious that I was approaching it: and I must have somehow crawled or won myself forward. I have a certain sort of impression that it was a purple land of pure porphyry; there is some faint memory, or dream, of hearing a long-drawn rumor of breakers booming upon its rock: I do not know how I have them. I certainly remember retching with desperate jerks of my travailing entrails, remember that I was on my back when I moved the adjustor in the engine-room: but any recollection of going down the stair, or of coming up again, I have not. Happily, the rudder being fixed hard to starboard, the ship, as she forged ahead, must have swung about; and I must have been back up to free the wheel in time, for when my senses came again I was lying there, my head against a gimbal, one heel stuck up on a spoke of the wheel, no land in sight, and the sun shining.

This made me so sick, that for either two or three days I lay without eating in the seat near the wheel, only waking occasionally to sufficient sense to see to it that she was making westward from that place; and on the morning when I came well to myself I was not certain whether it was the second or the third morning: so that my calendar, so exactly kept, may

Visit to rotten ship -

now be a day out, for to this day I have never been at the pains to ascertain if I am here spelling on the 10th or the 11th of May.

* * *

Well, on the fifth evening after this, as the sun was sinking at the rim of the sea, I happened to look where he hung on the starboard bow: and there I saw a black-green spot clean-cut against his red—a very unusual object here and now—a ship: a poor thing, as she proved when I got nigh to her, without any sign of mast, all water-logged, some relics of rigging straggling over her beam, even her bowsprit broken at the middle, she nothing but one bush of weeds and sea-things from bowsprit-tip to poop-edge, stout as a hedgehog, awaiting there the next pounding of the sea to founder.

It being near my dinner-hour, I stopped the *Speranza* about sixty feet from her; and, in pacing my spacious poop, as usual before eating, kept giving glances at her, wondering who were the sons of men that had lived on her, their names, and minds, and way of life, and faces, until the desire arose within me to go to her and see: so I threw off my outer robes, uncovered and unroped the cedar cutter—the only boat, except the air-pinnace, then left to me whole—and lowered her by the mizzen pulley-system. But it was a ridiculous nonsense, for when I had paddled to the derelict it was only to be thrown into paroxysms of rage by repeated failures to scale her bulwarks, low as they were: for though my hands could easily reach, I could find no hold on the slimy mass, and three rope-ends which I seized were also untenably slippery, so that I collapsed always back into the boat, my clothes a mass of filth, and the only thought in my blazing brain a twenty-pound charge of guncotton, of which I had plenty, to blast her backside to uttermost Hell. In the end I had to go back to the *Speranza*, get rope, then back to the other, for I would not be challenged in such a way, though now the dark was come, hardly tempered by a far half-moon, and I getting hungry, and from minute to minute more devilishly ferocious; until, by dint of throwing, I managed to slip the rope-loop round a mast-stump and drew myself up, my left hand slashed by some hellish shell: and for what? the imperiousness of a caprice. The shadowy moonshine shewed an ample tract of deck, mostly invisible beneath rolls of putrid seaweed, and no bodies, nothing but a concave esplanade of seaweed, she a ship of probably 3,000 tons, three-masted, a sailer. When I moved aft, having on thick babooshes, I could see that only four of the companion-steps remained; but by a leap I was able to descend into that desolation, where the stale sea-stench seemed concentrated into the very essence of rawness, and here I got a ghostly awe and timorousness, lest she should go down with me, or something; but, on flashing matches, I saw an ordinary cabin, with some fungoids, skulls, bones, rags, but not one connected skeleton; in the second starboard berth a table, and on the floor an ink-pot whose continual rolling made me look down: and there I observed a scribbling-book with black covers which curved half-open, for it had been wet. This book I took, and paddled back to the *Speranza*: for that ship was nothing but an emptiness, and a strench of the crude elements of existence, nearly assimilated already to the rank deep to which she was wedded, soon to be sucked back into its nature and being, to become a sea-in-little, as I, in time, my God, am to be turned into an earth-in-little.

During dinner, and after, I read the book—with some difficulty, for it was pen-written in French, and discolored; and it turned out to be the journal of someone, a passenger and voyager, I imagine, who called himself Albert Tissu, and the ship the *Marie Meyer*: nothing remarkable in the narrative—descriptions of South Sea scenes, records of weather, cargoes—until I came to the last page, which was remarkable enough, that page being dated the 12th of April—strange thing, my good God, that same day, twenty years ago, when I reached the Pole; and the writing on that page was quite different from the spruce look of the rest, proving high excitement, wildest haste, headed *"Cinq Heures, P.M.,"* and he writes: "Monstrous event! phenomenon without likeness! the witnesses of which must live immortalized in the annals of the universe, so that Mama and Juliette will now confess that I was justified in undertaking this voyage. Conversing with Captain Tombarel on the stern, when a murmur from him—'Mon Dieu!' His visage blanches! I follow the direction of his gaze to eastward—I behold! seven kilomètres perhaps away, *ten waterspouts,* reaching up, up, high, all in a line, with intervals of nine hundred mètres, very regularly placed; but they do not wander nor waver, as waterspouts do, nor are they at all lily-shaped, like waterspouts—just pillars of water, a little twisted here and there, and, as I conjecture, fifty mètres in diameter. And six minutes we look, while Captain Tombarel repeats and repeats under his breath 'Mon Dieu!' 'Mon Dieu!', the while crew now on deck, I agitated, yet collected, watch in hand; until suddenly all is blotted out: the pillars, doubtless still there, can no more be seen: for the ocean about them is steaming, hissing higher still than the pillars a vapor, immense in extent, whose sibilation we at this distance can distinctly hear. It is affrighting! it is intolerable! the eyes can hardly bear to watch, the ears to hear! it seems unearthly travail, monstrous birth! But it lasts not long: all at once the *Marie Meyer* commences to pitch and roll, for the sea, a moment since still, is now rough! and at the same time, through the white vapor, we descry a shade rising, a shade, a mighty back, a new-born land, bearing skyward ten flames of fire, slowly, steadily, out of the sea, into the clouds. At the moment when that sublime emergence ceases, or seems to cease, the thought that smites me is this: 'I, Albert Tissu, am immortalized': I rush down, I write it. The latitude is 16° 21′ 13″ South; the longitude 176° 58′ 19″ West[1]. There is a running about on the decks—an odor like almonds—it is so dark, I——"

So this Albert Tissu.

* * *

With all that region I woud have no more to do: for all here, it used to be said, lies a sunken continent, and I thought that it would be rising and shewing itself to my eyes, and driving me rushing-frenzied: for the earth is turbid with these contortions, monstrous grimaces, apparitions that are like the Gorgon's face, appalling a man into spinning stone; and nothing could be more appallingly insecure than living on a planet.

Nor did I stop until I had got so far north as the Philippines, where I was two weeks—exuberant, odorous places, but so steep and rude, that at one place I abandoned all attempt at travelling in the motor, and left it in a valley by a broad, noisy river, thick with mossy rocks: for I said "Here I will live, and be at peace"; and then I had a scare, seeing that during three days I could not re-discover the river and the motor, and I was in the greatest despair, thinking "When shall I find my way out of these jungles and vastnesses?", for I was where no paths are, and had lost myself in depths of verdure where the lure of the earth is too strong and rank for a solitary man, since in such places, I assume, a man would rapidly be transmuted into a tree, or a snake, or a cat. At last, however, I refound the spot, to my great joy, but would not shew that I was glad, and, to conceal it, attacked a wheel of the motor with some kicks. . . . But those two years of roaming, they are over, and like a dream; and not to write of that—of all that—have I taken this pencil in hand after seventeen long, long years.

* * *

Singular—my reluctance to put it on paper . . .

I will write of the voyage to China, how I landed the car on the wharf at Tientsin, and passed up, nigh the river, to Pekin through a maize and rice land which was charming in spite of cold, I thick with clothes like an Arctic traveller; and of the three earthquakes within two weeks; and how the only map which I had of the city gave no indication of the whereabouts of its military stores, and I had to seek them; and of the three days' effort to enter to them, every gate grim and riveted against me; and how I burned, but had to observe the flames from beyond the city-walls, the place being all one cursed plain; yet how I cried aloud with wild banterings and chal-

---

[1]French reckoning apparently, from meridian of Paris.

lenges of Tophet to that old Chinaman still alive within it; and how I coasted, and made acquaintance with the hairy Ainus, male and female hairy alike; and how, lying one midnight sleepless in my cabin, the *Speranza* being in a still glassy harbor beneath a cliff overgrown by drooping greenery—the harbor of Chemulpo—to me lying awake came the notion, "Suppose now you should hear a foot pacing to and fro, patiently, on the poop above—*suppose*"; and the night of terrors that rived me: for I could not help supposing, and at one time really seemed to hear it, and how sweat poured from my every pore; and how I went to Nagasaki, and destroyed it; and how I crossed that Pacific deep to San Francisco, for I knew that Chinamen had been there, too, and one might be alive; and how, one still day, the 15th or the 16th of April, I, seated by the wheel in the mid-Pacific, suddenly noted a wild white hole that ran and wheeled, and wheeled and ran, within the sea, reeling toward me; and I was aware of the hot whiff of a wind, then of the hot wind itself, which wheeled, deep-venting a vehemence of the letter *V*, humming the hymn of hosts of spinning-tops, and the *Speranza* was on her beam, sea pouring over her port-bulwarks, myself down on the deck against the taffrail, drowning fast, pegged there; but all was soon over, and the hole within the sea, and the hot spinning-top of wind, ran reeling on to the horizon, and the *Speranza* righted herself: so that it was evident that someone wished to do for me, for that a typhoon of such vehemence ever blew before I do not think; and how I arrived at San Francisco, and fired it, and had my delights: for it was mine; and how I thought to pass over the trans-continental railway to New York, but would not, fearing to leave the *Speranza*, lest all the craft in the harbor there should be wrecked, or rusted, and buried under sea-weed, and turned unto the sea; and how I returned, my thoughts all seduced now to musing on the earth and her moods, and a notion in my soul that I would return to those secret deeps of the Filipinas, and evolve into an autochthone—a sycamore or a serpent, or a person with serpent-limbs, like the Saturnian autochthones; but I would not: for Heaven was in men, too, Earth and Heaven; and how, as I moved round west anew, another winter come, I now lost in a mood of dismal despondencies, on the very brink of the inane abysm and smiling idiotcy, I saw in the island of Java that temple of Boro Budor: and like a tornado, or volcanic event, my soul was changed: for my studies in the architecture of man before I started on the palace came back to me with zest, and for five nights I slept in the temple, examining it by day. It is vast, having that aspect of massiveness which characterizes Mongol building, my measurement of its breadth being 529 feet, and it rises in six terraces, each divided up into innumerable niches, containing each a statue of the seated Boodh, with a voluptuousness of tracery that is drunkening, all surmounted by a crowd

Back at Imbros

of cupolas, and crowned by a great dagop: and when I saw this, I had a longing to be back at my home after so prolonged roaming, and to set up the temple of temples; and I said: "I will go back, and build it as a witness to God."

* * *

Save for some days in Egypt, I did not once stop on that homeward voyage, moving into the little harbor at Imbros on a calm sundown on the 7th of March (as I reckon); and I moored the *Speranza* to the ring in the little quay; then raised the battered motor from the hold with the middle air-engine ("battered" by the typhoon in the mid-Pacific, which had broken it from its ropings and tumbled it head-over-heels to port); and I went through the windowless village-street, and up through the plantains and cypresses which I knew, and the Nile-mimosas, and mulberries, and Trebizond-palms, and pines, and acacias, and fig-trees, until thicket stopped me, and I had to alight, for in those two years the path had finally disappeared; and on, on foot, I made my way, until I came to the board-bridge, and leant there, and looked at the rill; and thence walked toiling up the foot-path in the sward toward that rolling table-land where I had built with many a groan, until, half-way up, I saw the tip of the crane-arm, then the blazing top of the south pillar, then the shed-roof, then the platform, a wobbling blotch of brightness watering the eyes under the setting sun. But the tent, and almost all that it had contained, was gone.

* * *

For two days I would do nothing, just lounging and watching, shirking a load so huge; but on the third morning I languidly began something: and I had not worked an hour when a fervour took me—to finish it, to finish it—and this did not leave me, with but three brief intervals, for nearly seven years; nor would the end have been so long in coming, but for the unexpected difficulty in getting the four flat roofs water-tight, I having to take down half the west one. Finally I made them of gold slabs 1¼ inch thick, on each beam double-gutters being fixed along each side of the top flange to catch any leakage at the joints, which are filled with slater's cement, the slabs being clamped to the top flanges by steel clips, having bolts set with plaster-of-Paris in holes drilled in the slabs, and the roofs are slightly pitched to the front edges, where they drain into gold-plated copper-gutters on plated brackets, with one side flashed up. . . . But now I babble again of that slavery, which I would forget, but cannot: for every measurement, bolt, ring, is in my brain, like an obsession; but it is past—and it was vanity.

* * *

Six months ago today it was finished: six months more protracted, desolate, burdened, than all those sixteen years in which I built.

I wonder what a man—some Shah, or Tsar, of that far-off past—would say now of me, if an eye could light on me? He'd shrink, I think—yes, undoubtedly—before the majesty of these eyes; and though I am not lunatic—for I am not, I am not—no doubt he'd fly from me, crying out "Here is the lunacy of Pride!"

For there would seem to him—I believe so—in myself, in all about me, somewhat of royal beyond bounds, fraught with terror. My body has fattened, my girth now filling out to a portly roundness its band or girdle of crimson cloth a foot broad, Babylonish, gold-embroidered, hung with a hundred copper and gold coins of the Orient; my beard, still ink-black, sweeps in two sheaves to my hips, flustered by each wind; as I pace the chambers of this palace, the floor of amber-and-silver blushes in its depths, reflecting the low neck and short arm of my robe of blue and scarlet, abloom with luminous stones. I am ten times Satrap and Emperor, seated a hundred times enthroned in established obese old majesty: challenge me who dare! Among those lights that I nightly pore upon may fly songsters, my peers and fellow-denizens, but *here* I am sole; earth bows her brow before my purples and hereditary sceptre: for though she entices me, not yet am I hers, but she is mine. It seems to me not less than a million æons since other beings, more or less resembling me, stepped impudently in the open sunlight on this planet; I can in fact no longer picture to myself, nor properly credit, that such a state of things—so fantastic, far-fetched, droll—could have existed: though, at bottom, I suppose, I know that it must have been so; indeed, up to ten years ago I used to *dream* that there were others, would see them go about the streets like ghosts, and be troubled, and bound awake; but never now could such a thing, I think, occur to me in sleep: for the wildness of the circumstance would certainly strike my mind, and immediately I should descry that the dream was a dream. For now at least I am sole, I am lord. The walls of this palace which I have piled stare down ravished at their reflection in the fire of a lake of wine.

Not that I made it of wine because wine is rare, nor the walls of gold because gold is rare, since I am not a goose: but because, having determined to match for beauty a human work with the works of those Others, I had in mind that, by some prank of the earth, precisely the objects most costly are usually the most beautiful.

The vision of splendor and loveliness which is this palace now risen before my eyes cannot be described by pen or paper, though there *may* be words in the lexicons of mankind which, if I searched for them with inspired wit for sixteen years, as I have built for sixteen years, might as vividly express my mind

Descrip. of his temple

to a mind as the stones-of-gold, so grouped, express it to the eye: but, failing such labors and skill, I suppose I could not give, if there lived another man, and I sought to give, the smallest conception of its celestial charm.

It is a structure not less clear than the sun, nor fair than the moon—the sole structure in the making of which no restraining thought of cost has played a part, one of its steps being of more cost than all the temples, mosques and besestins, the palaces, pagodas and cathedrals, reared between the eras of the Nimrods and the Napoleons.

The house itself is quite small—40 ft. long, by 35 broad, by 27 high: yet the structure as a whole is pretty enormous, high uplifted, because of the platform on which the house stands, its base 480 ft. square, its height 130 ft., its top 48 ft. square, the elevation 22½ degress, the top reached on each edge by 183 steps, low, gold-plated—not a continuous flight, but broken into threes, fives, sixes, nines, with landings between, these from the top looking like a great terraced parterre of gold: the palace is thus Assyrian in plan, except that the platform has steps every-way, instead of one set, the platform-top round the house being a mosaic of squares of the glassiest gold and of the glassiest jet, corner to corner, each square 2 ft., round the platform running 48 gold pilasters, 2 ft. high, square, tapering upwards, topped by knobs, the knobs connected by silver chains, from the chains hanging hosts of silver globes that gabble together in a breeze. The house itself consists of an outer court (facing east toward the sea) and the house proper built round an inner court, the outer court being an oblong as broad as the house, its three walls of gold, battlemented, lower than the house, round their top running a band of silver 1 ft. wide; and at the gate, which is Egyptian, narrower at top, stand the two pillars of gold, square, tapering upwards, 45 ft. high, with their capital of band, closed-lotus, and plinth. In the outer court is the well, reproducing in little the shape of the court, its sides gold-lined, tapering downwards to the bottom of the platform, where a conduit replenishes the mean evaporation of the lake—automatically on the principle of carburetter-floats—the well containing 105,360 litres, and the lake occupying a circle round the platform of 980 ft. diameter, with a depth of 3½ ft. Round the well, too, run pilasters connected by silver chains, and it communicates by a conduit with a pool of wine sunk into the inner court, the pool being fed from eight gold tanks, tall and narrow, tapering upwards, which surround it, each containing a different red wine, sufficient to last my lifetime. The ground of the outer court, as well as the platform-top, is a mosaic of jet and gold, but thenceforth the squares consist of silver and amber, amber limpid as slabs of solid oil, the entrance to the inner court being by an Egyptian doorway with folding-doors of cedar, gold-plated, surrounded by a coping of silver, huge, thick, 3½ ft. wide, simplicity of line everywhere heightening the effect of richness of material. The rest resembles rather a Homeric than an Assyrian house (except for the "galleries", which are Babylonish and Old Hebrew), the inner court with its wine-pool and tanks being an oblong 8 ft. by 9 ft., upon which open four silver-latticed windows, oblongs in the same proportion, and two doors, oblongs in the same proportion, round this court running the eight walls of the house proper, the four inner being 10 ft. from the four outer, each parellel two forming one long chamber, except the front (east) two, which are split up into three rooms. In each room are four panels of silver, thinner than their rims, in the sunken space being pantings, of which 21 were taken at the burning of Paris from a place named "The Louvre," and 3 from a place in London, the panels having the look of great frames, and are surrounded by garlands of opal, garnet, topaz, each garland being an oval, a foot wide at the sides, narrowing to an inch at top and bottom. As to the "galleries," they are four recesses in the four outer walls under the roofs, hung with rose and white silks on gold pilasters, each gallery entered by four steps down from its roof, to the roofs leading two corkscrew stairs of cedar, east and north, on the east roof being the kiosk with the telescope: and from that height, and from the galleries, I can watch under the moonlight of this climate, which is not unlike limelight, those mountains of Macedonia silent for ever, and where the islands of Samothraki, Lemnos, Tenedos sleep like purplish birds of fable on the Ægean Sea: for, usually, I sleep during the daytime and keep a nightlong vigil, frequently at midnight descending to be taking my scarlet baths in the lake, to disport myself in that intoxication of nose, eyes, pores, dreaming long wide-eyed dreams at the bottom, to come back up doddering, weak, drunken. Or again—*twice* within these idle void months—I have rushed, calling out, from these halls of luxury, snatching off my gorgeous rags, to skulk in a hut on the shore, smitten in those moments with a vision of the past and vastness of this planet, and moaning "alone, alone . . . all alone, alone, alone . . . alone, alone. . . ." for events resembling eruptions take place in my brain, and one flushed 'foreday—how flushed!—I may kneel on the roof with streaming cheeks, my arms cast out, with awe-struck heart adoring, the next I may strut like a cock, wanton as sin, lusting to blow up a city, to wallow in filth, and, like the Babylonian maniac, naming myself the mate of Heaven.

* * *

But it was not to write of this—of all this . . .

Of the furnishing of the palace I have written nothing. . . . But why I hesitate to admit to myself what I *know* . . . If They speak to me, I may speak of Them: for I do not fear Them, but am Their peer. . . .

Of the island I have written nothing: its size, climate, form, flora. . . . There are two winds: a north and a south; the north is cool, the south is warm; and the south blows during the winter months, so that sometimes at Christmas it is hot; and the north blows from May to September, so that the summer is seldom oppressive, and the climate was made for a king. The mangal-stove in the south hall I have never once lit.

The length is 19 miles, the breadth 10, and the highest mountains must be 2,000 ft., though I have not been all over it. It is densely wooded, and I have seen growths of wheat and barley, obviously degenerate now, with currants, figs, valonia, tobacco, vines in rank abundance, and two marble quarries. From the palace, which stands on a sunny plateau of swards, dotted with the shades thrown by fourteen huge cedars and eight planes, I can see all round an edge of forest, with the sheen of a lake to the north, and in the hollow to the east the rivulet with its bridge; and I can spy right through——

* * *

It shall be written now:

I have this day heard within me the contention of the voices.

* * *

I had thought that they were done with me! That all, all, all, was ended! I have not heard them for twenty years!

But today—distinctly—breaking in with brawling suddenness upon my consciousness . . . I heard.

This *far niente* and vacuous inaction here has been undermining my mind, this brooding upon the earth, this empty life, and bursting brain! So, immediately after eating at noon today, I said to myself "I have been duped by the palace: for I have spent myself in building, hoping for peace, and there is no peace; therefore now I will flee from it to another, sweeter work—not of building, but of burning—not of Heaven, but of Hell—not of self-denial, but of reddest revel: Constantinople—beware!"; and, throwing a plate away, with a stamp I was up: but, as I stood—again, again—I heard: the startling wrangle, the vulgar rough outbreak and voluble controversy, till my consciousness could not hear its ears; and one urged: "Go! go!" and the other: "Not there! . . . where you like . . . not there . . . for your life!"

I did not—for I could not—go, I was so overcome; dropped shuddering upon the couch.

These voices, or impulses, strongly as I was conscious of them of old, quarrel within me now with an openness new to them. Lately, influenced by my scientific habit, I have asked myself whether what I used to call "the Voices" were not in truth two intuitive movements such as most men may have felt, though with less force. But today doubt is past, doubt is past: nor, unless I be mad, can I ever more doubt.

* * *

I have been thinking, thinking, of my life: there is a something which I cannot understand.

There was a man whom I met in that dark backward and abysm of time—at the college in England it was—his name far enough now beyond the grasp of my memory, lost in the limbo of past things; but he used to talk about certain "Black" and "White" Powers, and of their strife for this world—short man with a Roman nose, who lived in fear of growing a paunch, his forehead in profile more prominent at top than at bottom, his hair parted in the middle, and he had the theory that the male form was more beautiful than the female—I forget what his name was, the dim clear-obscure being, one of those untrained brains that accepted fancies and ascertained facts with equal belief, as men in general did: yet deep was the effect of his thesis upon me, though I think I often made a point of mocking him. This man always declared that "the Black" would carry off the victory in the end: and so he has, old "Black."

But, assuming the existence of this "Black" and this "White" being—and supposing it to be a fact that my reaching the Pole had any connection with the destruction of species, according to the notions of the extraordinary Scotch "preacher"—then, it must have been the potency of *"the Black"* which carried me, over all obstacles, to the Pole. So far I can understand.

But *after* I had reached the Pole, what further use had either White or Black for me? Which was it—White or Black—that preserved my life through my protracted return on the ice—and *why?* It *could* not have been "the Black!" For from the moment when I stood at the Pole, the only purpose of the Black, which had formerly preserved, must have been to destroy, me with the others. It must have been *"the White,"* then, that led me back, retarding me long, so that I should not enter the poison-cloud, and then openly presenting me the *Boreal* to bring me home to Europe. But his motive? And the significance of these fresh wrangles, after such a stillness? This I do not understand!

Damn Them and their tangles! I care nothing for Them!—if they were there. For are not these outcries that I hear nothing but the screams of my own burning nerves, and I all mad and morbid, morbid, and mad, mad, my good God?

This inertia here is *not good* for me! This stalking about the palace! and long thinkings about Earth and Heaven, Black and White, White and Black, and things beyond the stars! My brain is like bursting through the walls of my poor head.

Tomorrow, then, to Constantinople . . .

* * *

At Constantinople To burn it -

I came down to the *Speranza* with the motor, went through her, spent the day in work, slept on her, worked again today until four at both ship and time-fuses (I with only 700 fuses left, and in Stamboul alone must be 8,000 houses, without counting Galata, Tophana, Kassim-pacha), started out at 5.30, and am now at 11 p.m. lying two miles off the island of Mamora, with moonlight musing on the sea, which a breeze brindles, the little land seeming immensely stretched-out, grave and great, as if it were the globe, and there were nothing more, and the tiny island at its end immense, and the *Speranza* vast, and I alone puny. Tomorrow morning I will moor the *Speranza* in the Golden Horn at that hill where the palace of the Capitan Pacha is. . . .

* * *

I found that tangle of craft in the Golden Horn wonderfully preserved, with hardly any moss-growths, owing, I suppose, to the little Ali-Bey, which, flowing into the Horn at the top, makes a constant current. . . .

Ah, I remember the place; long ago I lived here—the fairest of cities—and the greatest, for, though I think that London in England was bigger, no city, surely, ever *seemed* so big. But it is flimsy, and will burn like tinder, the houses built light, of timber, with interstices filled by earth and bricks, some looking ruinous already, with their lovely tints of green and gold and pink and azure and daffodil, faint like tints of flowers withering: for it is a city of paints and trees, and all about the little winding streets, as I write, are volatile armies of almond-blossoms, laughing in a mêlée with maple-blossoms, white whirled with purple. Even the most sumptuous of the Sultan's palaces are built in this combustible manner, for I believe that they had a notion that stone-building was presumptuous, though I have seen some stone-houses in Galata; indeed, the place lived in a state of sensation at nightly flares-up, and I have come across several tracts already devastated by fires. The ministers-of-state used to attend them, and, if the fire would not go out, the Sultan himself would drive up, to egg-on and incite the firemen. Now it will burn still better.

But I have been here six weeks, and still no burning: for the place seems to plead with me, it is so fair, and I do not know why I did not live here, and spare my toils all those sixteen years of nightmare: so that for three weeks the impulse to fire was quieted, and since then an irritating whisper has been at my ear which says: "It is not really like the Shah you are, this firing, rather like a child, or a savage, who liked to see fireworks; at least, if you must burn, do not burn poor Constantinople, which is so charming, and so old, with its balsamic perfumes, and the blossomy trees of white and light-purple peering over the walls of the cloistered houses, and all those lichened tombs—menhirs and regions of marble tombs between the quarters, Greek tombs, Byzantine, Jew, Mussulman tombs with their strange and sacred inscriptions, overwaved by their cypresses sighing, and their plane-trees"; and for weeks I would do nothing, but roamed about with two minds in me under the sultriness of the sky by day, and the mighty trance of the nights of this place, that are like nights gazed at through azure glasses, and in one of them is not one night, but the thousand-and-one crowded nightlongs of glamour and phantasm: for I would sit on that esplanade of the Seraskierat, or those tremendous stones of the porch of the mosque of Mehmedfatih, dominating from its steps all Stamboul, and pore upon the moon for hours and hours, so passionately rapt she soared through cloud and cloudless, until I would be smitten with doubt of my own identity: for whether I were she, or the earth, or myself, or some other thing or person, I did not know, all so silent alike, and all, except myself, so vast, the Seraskierat, and Stamboul, and the Marmora Sea, and Europe, and those argent fields of the moon, all large alike compared with me, and measure and space were lost, and I with them.

* * *

These proud Turks died stolidly, many of them: in streets of Kassim-pacha, in crowded Taxim on the heights of Pera, and under the arcades of Sultan-Selim, I have seen the open-air barber's razor with his bones, and with him the skull of the faithful half-shaved, and the two-hours' narghile with traces of tembaki and haschish still in the bowl. Ashes now are they, and dry yellow bone; but in the houses of Phanar, in noisy old Galata, in the Jew quarter of Pri-pacha, the black shoe and head-dress of the Greek is still distinguishable from the Hebrew blue: for it was a ritual of colors here in boot and hat—yellow for Mussulman, red boot, black calpac for Armenian, for the Effendi a white turban, for the Greek a black, while the Tartar skull shines from under a high calpac, the Nizain-djid's from a melon-shaped head-piece, the Imam's and Dervish's from a conical felt, and here and there a "Frank" in European rags; and I have seen the towering turban of the bashi-bazouk, and some softas in those domes on the wall of Stamboul, and the beggar, and the street-merchant with his tray of watermelons, sweetmeats, raisins, sherbet, and the bear-shewer, and the Barbary organ, and the night-watchman, who evermore cried "Fire!", with his lantern, pistols, dirk, and wooden javelin; I have gone out to those plains beyond the walls whence the city looks nothing but minarets shooting through cypress-tops, and I seemed to see the muezzin at some summit, crying "*Mohammed Resoul Allah!*"—the wild man; and from the cemetery of Scutari the walled city of Stamboul lay spread entire before me up to Phanar and Eyoub in their cypress-woods, the whole embowered now, one mass of alleys darkened by balconies of old Byzantine houses, beneath which one on mule-back had to stoop the head—alleys where even old Stamboulers would lose their way in intricacies of the picturesque; and within the boscage of the Bosphorus coast, to Foundoucli and beyond, some peeping yali, snow-white palace, or Armenian cot; and the Seraglio by the sea, a town within a town; and southward the sea of Marmora, blue-and-white, and vast, wriggling vigorous like a sea just born and rejoicing at its birth under the sun, all brisk, alert, to the islands like sighs afar: and, as I looked, I suddenly said a wild, mad thing, my God, a wild and maniac thing, a screaming maniac thing for Hell to scream at: for something said with my tongue: *"This city is not quite dead."*

* * *
* * *

Five nights I slept in Stamboul itself at the palace of some sanjak-bey or emir, or rather dozed, with one slumbrous lid that would open to note my visitors Sinbad, and Ali Baba, and old Haroun, to note how they slumbered and dozed: for it was in the small chamber where the bey received those speechless all-night visits of the Turks, rosy hours of perfumed romance, and drunkenness of the fancy, and visionary languor, sinking toward sunrise into the still deeper peace of sleep; and there, still, were the *yatags* for the guests to sit cross-legged on for the waking mooning, and to drop upon for the morning swoon, and the copper brazier still scenting of essence-of-rose, and the cushions, rugs, hangings, the monsters of the wall, the haschish-chibouques, hookahs, narghiles, and drugged pale cigarettes, and a secret-looking lattice outside the doorway, painted with trees and peacocks; and the air narcotic and grey with the incense of pastilles and the scented smokes that I had smoked; and I all drugged and mumbling, my left eye suspicious of Ali there, and Sinbad, and old Haroun, who dozed. And when I had slept, and rose to bathe in a room close to the latticed balcony of the façade, before me lay Galata in sunshine, and that great avenue mounting to Pera, once crowded at every nightfall with divans on which grave dervishes smoked narghiles, and there was no room to pass, for all was divans, lounges, almond-trees, heaven-high hum, chibouques in forests, the dervish, and the innumerable porter, the horse-hirer with his horse from Tophana, and arsenal-men from Kassim, and traders from Galata, and artillery-workmen from Tophana; and at the back of the house a covered bridge led across a street, which consisted of two walls, into a wilderness of flowers, all a tangle, which was the harem-garden, where I passed some hours; and here I might have remained many days, but that dozing one 'foreday with those fancied others, it was as if there occurred a laugh somewhere, and a thing said: "But this city is not quite dead!" startling me from deeps of peace to wakefulness; and I said to myself: "If it is not quite dead, it *will* be—with some suddenness!": and that morning I was at the Arsenal.

* * *

It is long since I have so enjoyed, to the spine. It may be "the White" who has the guidance of my life, but assuredly it is "the Black" who governs in my soul.

Grandly did old Stamboul, Galata, Tophana, Kassim, right out beyond the walls of Phanar and Eyoub, blaze and flare—the whole place, except one bit of Galata, being like so much tinder: and in the five hours between 8 p.m. and 1 a.m. all was over. I saw the tops of all that forest of cemetery-cypresses round the tombs of the Osmanlis outside the walls, and those in the cemetery of Kassim, and those round the mosque of Eyoub, shrivel away instantaneously, like flimsy hair snatched by a flame; I saw the Genoese tower of Galata go heading obliquely on an upward curve, like Sir Roger de Coverley and wild rockets, and burst high, with a report; in pairs, and trios, and fours, I saw the cupolas of the fourteen great mosques give in and sink, or soar and rain, and the great minarets nod the forehead, and drop; and I saw the flame-sheets reach out and out across the empty breadth of the Etmeidan—three hundred yards—to the six minarets of the Mosque of Achmet, wrapping the red-granite obelisk in the center; and across the breadth of the Serai-Meidani it reached to the building of the Seraglio and the Sublime Porte; and across those waste spaces between the houses and the great wall; and across the seventy or eighty arcaded bazaars, all-enwrapping, it reached; and the spirit of fire grew upon me: for the Golden Horn itself was a tongue of fire, crowded, west of the galley-harbor, with exploding battleships, corvettes, frigates, brigs, and, east, with a region of gondolas, feluccas, caïques, merchantmen, aburn; on my left crackled Scutari; and I had sent out forty craft under low horse-powers of air, with fuses timed for 11 p.m., to light with their roaming fires the Sea of Marmora: so before midnight I was girdled in one furnace and gulf of fire, sea and sky inflamed, and earth aflare. Not far from me to the left I saw the Tophana barracks of the Cannoniers, and the Artillery-works, after long reluctance and delay, take wing together; and three minutes later, down by the water, the barrack of the Bombardiers and the Military School together, grandly, grandly; and then, to the right, in the valley of Kassim, the Arsenal: these five riding the sky like smoky suns, and pouring daylight of Tophet over many a mile of sea and country; also I saw the two lines of ruddier flaring where the barge-bridge and the raft-bridge over the Golden Horn galloped in haste to burn; and all that vastness burned in haste, faster and faster—to fervor—to carnival—to unanimous acme: and when its roaring railed at the infinite, and the might of its glowing heart was gravitation, being, sensation, and I its compliant wife, then my forehead dropped, and, sighing as it were my final sigh, I tumbled drunk.

* * *
* * *

O wild Providence! Unfathomable madness of Heaven! that ever I should write what now I write! I will not write it. . . .

* * *

The hissing of it! It must be some frantic fancy! a tearing out of the hair to scatter on the ranting fire-cataracts of Saturn! My hand will not write it!

* * *

In God's name . . . During four nights after the fire I slept in a house—French, as I saw by the books, &c., probably the Ambassador's, for it has vast gardens and a good view over the sea, situated on that east declivity of Pera—one of the houses which, for my safety, I had left standing round the minaret whence I had watched, this minaret being at the top of the Mussulman quarter on the heights of Taxim, between Pera proper and Foundoucli; and down below, both at the quay of Foundoucli and at that of Tophana, I had left under shelter two caïques for double-safety, one a Sultan's gilt craft, with the gold-spur at the prow, and one a boat of those zaptias that patrolled the Golden Horn as water-police: by one or other of which I meant to reach the *Speranza,* she being safely anchored some distance up the Bosphorus coast. So on the fifth morning I set out for the Tophana quay; but, as some rain had fallen overnight, this had re-excited the thin smoke resembling quenched steam, which, as from some reeking district of Abaddon, still trickled upward over many a square-mile of blackened tract, through of flame I could see no sign; and I had not advanced far over every sort of *débris* when I found my eyes watering, my throat choked, my way almost blocked by roughness: whereupon I said "I will turn back, cross the region of tombs and barren behind Pera, descend the hill, get the zaptia boat at the Foundoucli quay, and so reach the *Speranza.*"

Accordingly, I made my way out of the quarter of smoke, walked beyond the limits of smouldering ruin and tomb, and soon entered a woodland, singed at the beginning, but soon green and flourishing as the jungle. This cooled and soothed me; and, being in no hurry to reach the ship, I was led on and on, in a north-western direction, I think. Somewhere thereabouts, I thought, was the place they called "The Sweet Waters," and I went on with some notion of coming upon them, thinking to pass the day, until afternoon, lost in that forest, where nature in just twenty years has rushed back to an exuberance of savagery, everywhere now the wildest vegetation, dim dells, rills wimpling through twilights of mimosa, pendulous fuchsia, palm, cypress, mulberry, jonquil, narcissus, daffodil, rhododendron, acacia, fig. Once I stumbled upon a cemetery of old gilt tombs, absolutely overgrown and lost, and anon got glimpses of little trellised yalis choked in boscage, as with a listless foot I moved, munching an almond or an olive, though I could vow that olives were not formerly indigenous to any soil so northern; yet here they are now in plenty, though elementary: so that modifications whose end I cannot see are clearly proceeding in everything, some of the cedars that I met that day being immense beyond anything I ever saw; and the thought, I remember, was in my head, that if a twig or a leaf should turn into a bird, or into a fish with wings, and fly before my eyes, what then should I do? and I would eye a bush suspiciously a little. After a long time I penetrated into a very sombre grove, where, the day outside the wood being brilliant, grilling, breathless, the leaves and flowers hung motionless, so that I seemed to be hearing on my ear-drum the booming of the muteness of the universe, and when my foot split a twig it produced the report of pistols. Then I got to a glade in the tangle, about eight yards across, that gave out a fragrance of lime and orange, where the twilight just enabled me to see some old bones, three skulls, the edge of a tam-tam prying out from a tuft of wild corn with corn-flowers, some golden champac, and all round a gushing of muskroses. I had stopped—*why* I do not recollect—perhaps at the thought that, if I was not getting to the Sweet Waters, I should seriously be setting about seeking my way out; and, as I stood looking about me, I remember that some cruising insect drew near my ear its lonesome drone.

Suddenly, God knows, I started. . . .

I believed—I dreamed—that I saw a pressure in a bed of moss and violets, *recently made!* and while I stood poring upon that impossible thing, I believed—I dreamed—the lunacy of it!—that I heard a laugh . . . the laugh, my good God, of a human soul.

Or it seemed half a laugh, and half a sob: and it passed from me in one fleeting instant.

Laughs, and sobs, and absurd hallucinations, I had often heard before, feet walking, noises behind me; and, even as I had heard them, I had known that they were nothing: but, brief as was this impression, it was yet so thrillingly *real*, that my heart received as it were the shock of death, and I was shot backward into a mass of moss, where I remained sustained on my right palm, while the left pressed my laboring breast; and there, toiling to draw my breath, I lay still, all my soul focussed into my ears; but now could hear no sound, save only that hum of the dumbness of the inane.

He encounters a living female being

There was, however, the foot-print: if my eye and ear should so conspire against me, that, I thought, was hard.

Still I lay, still, in that same position, without a stir, sick and dry-mouthed, infirm, with dying breaths: but keen, keen —and malign.

I would wait, I said to myself, I would be cunning as snakes, though so woefully sick and invalid: I would make no sound. . . .

After some time I became aware that my eyes were leering —leering in one direction: and immediately the fact that I had a sense of direction proved to me that I must, *in truth*, have heard something! Whereupon I strove—I contrived—to raise myself; and, as I stood upright, swaying there, not the terrors of death alone were in my breast, but the authority of the monarch was on my forehead.

I moved: I found the strength. . . .

Slow step by slow step, with daintiest noiselessness, I moved to a thread of moss that led from the glade into the grove; and along its zigzag way I wound—toward the sound, in my ears now the noising of some streamlet, while, following the moss-path, I was led into a mass of bush which reached only two or three feet above my head; and through this, stealing, I wheedled my painful way, got out upon a strip of long-grass, to be faced now by a wall of acacia-trees, prickly-pear, pichulas, three yards before me: between which and forest beyond I got glimpses of a streamlet's gleams.

On my hands and knees I crept toward the acacia-thicket; entered it a little; and, leaning far forward, peered. And there —at once—ten yards in front, rather to my right—I saw.

Strange to say, my agitation, instead of intensifying to the point of apoplexy and death, now, at the actual sight, subsided to something like calm: and with a malign and sullen eye askance I knelt, eyeing her there.

* * *

She was on her knees, her palms on the ground supporting her, at the margin of the streamlet; leaning over she was, eyeing with a species of shyness, and of startled surprise, the reflection of her face in the waves: and I, with a sullen eye askance, knelt there, and finally stood, regarding her during five, six, good minutes of time.

* * *

I believe that her half-a-laugh and half-a-sob which I had heard had been the effect of astonishment at seeing her image in water; and I firmly believe, from the expression of her face, that this was the first day that she had seen it.

* * *

Never, I felt, as I observed her, had I beheld on earth a being so fair (though, analyzing now at leisure, I can conclude that in reality there was nothing very remarkable about her looks): her hair, fairer than auburn, and frizzy, forming a real robe to her nudity, robing her below the hips, some strings of it falling, too, into the water; her eyes, a violet blue, wide in the silliest look of bewilderment; and when, while I eyed and eyed her, she slowly rose, at once I remarked in all her manner an air of unfamiliarity with nature, as of one all at a loss what to do, her pupils looking unused and shy to light, and I could swear that that was the first day in which she had seen a tree or a stream.

Her age appeared seventeen or eighteen; I could conjecture that she was of Circassian blood, or, at least, origin; her skin whitey-brown, or old ivory-white.

* * *

Motionless she stood, at a loss: took a lock of her hair, and drew it through her lips; and there was some look in her eyes, which I could now plainly see, that somehow indicated a hunger going wild, though the wood was full of food. After letting go her hair, she stood again feckless and imbecile, her head hung sideward, pitiable to see I think now: for, though no faintest pity visited me then, it was evident that she did not know what to make of the look of things. At last she sat on a moss-bank, reached and took a musk-rose, put it on her palm, looked hopelessly at it.

* * *

One minute after my actual sight of her my excess of excitement, I say, had died down to something like calm. The earth was mine by old right: I felt that; and this creature a slave, upon whom, without heat or haste, I might perform my will: and for several minutes I stood coolly enough considering what that will should be.

The little canghiar, its silver handle encrusted with coral, its curved blade sharp as a razor, was as usual at my girdle: and the obscenest of the fiends was whispering at my ear with persistence: "Kill, kill—and eat."

Why I should have killed her I do not know: that question I now ask myself, wondering now whether it may be true, true, that it is *"not good"* for man to be alone. There was a religious sect in the past which called itself "Socialist," and with these must have been the truth, man being at his highest when most social, at his lowest when isolated: for the earth gets hold of all isolation, and draws it, to make it fierce, base, and materialistic, like sultans, aristocracies, and so on; but Heaven is where two or three are gathered together. It may be so: I do not know, nor care; but I know that after twenty years of loneliness on a planet the soul of man is more enamored of loneliness than of living, shrinking like a nerve

Bolt from sky stops AJ from killing woman

from the rude intrusion of another into the furtive realm of self, shrinking with that bitterness with which solitary castes —Brahmins, patricians, aristocracies, monopolists—always resisted any attempt to invade their domain of privileges. Also it may be true, it may, it may, that after twenty years of solitary selfishness a man becomes, without suspecting it, without noticing the stages of the evolution, a real and true beast, a Rome-burning Nero, a horrible, hideous beast, rabid, prowling, like that King of Babylon, his nails like birds' claws, his hair like eagles' feathers, with instincts all inflamed and fierce, delighting in darkness and crime for their own sake. I do not know, nor care; but I know that, as I drew the canghiar, the crookedest and the slyest of the guiles of the Pit was whispering me, tongue in cheek, "Kill, kill—and wallow."

With anguished gradualness, as a glacier stirs, tender as a nerve of each leaf that touched me, I moved, I stole, toward her through the belt of bush, the knife behind my back—steadily though slow—till there came a restraint, a check—I felt myself held back—had to stop—one of the sheaves of my beard having caught in a limb of prickly-pear.

I set to disentangling it; and it was, I believe, at the instant of succeeding that I first observed the condition of the sky, a strip of which I could spy across the rivulet: a sky which a little previously had been pretty clear, but now was busy with clouds; and it was a sinister muttering of thunder that had made me raise my lids, and see it.

When my eyes came down again to the sitting figure, she was looking foolishly round the sky with an expression which as good as proved that the girl had never before heard that sound of thunder, or, anyway, had no notion what it could bode: for my fixed leer lost not one of her actions, while, inch by inch, not breathing, cautious as the poise of a balance, I crawled. And suddenly, with a rush, I was out in the open, running her down. . . .

She leapt: perhaps two, perhaps three, paces she fled; then stock still she stood—within five yards of me—with expanded nostrils, with enquiring eyes.

I saw it all in one instant, and in one instant all was over. I had not checked the impetus of my run at her stoppage, and was on the point of reaching her with the knife uplifted, when I was checked and stricken by a stupendous violence: a flash of blinding light, attracted by the blade in my hand, struck jarring through my frame, and in the same moment the most passionate crash of thunder that ever racked a poor human heart felled me flat. The canghiar, snatched from my hand, pitched near the creature's feet.

I did not entirely lose consciousness, though surely, the Powers no longer hide themselves from me, and their contact is too intolerably rough and vigorous for a poor mortal man: so during, I think, three or four minutes I lay so astounded under that bullying outcry of wrath, that I could not budge an inch; and when at last I did sit up, the creature was standing near me with a kind of smile, holding out to me the weapon in a pouring rain.

I took it from her, and my doddering fingers dropped it into the stream.

* * *

Pour, pour, came the rain, raining as it can in this place, not long, but a torrent while it lasts, dripping in thick liquidity like a profuse sweat through the wood, I seeking to get back by the way I had come, fleeing, but with difficulty through the embarrassment of timber, and a feeling in me that I was being tracked—as it proved: for when I struck into more open space, almost opposite the west walls, but now on the north side of the Golden Horn, where there is a flat grassy ground somewhere between Kassim and Charkoi, with horror I saw that *protégée* of Heaven, or of someone, not twenty yards behind, following after me like a mechanical figure, it being now three in the afternoon, the rain drowning me through, I weary and hungry, and from all the ruins of Constantinople not one whorl of smoke going up.

I tramped on until I came to the quay of Foundoucli, and the zaptia boat; and there she was with me still, her hair nothing but a thin drowned string down her back.

* * *
* * *

Not only can she not speak to me in any language that I know, but she can speak in *no* language: it is my belief that she has *never* spoken; and she never saw a boat, or water, till now, I could vow.

She dared to come into the boat with me, sat clinging for dear life to the gunwale by her finger-nails, while I paddled the eight hundred yards to the *Speranza;* and she came up to the deck after me, astonishment imprinted on her face when she saw the open water, the boat, the yalis on the coast, and then the ship. But she appears to know little fear—smiled like a child, and on the ship touched this and that, as if each were a living thing.

When I went down to my cabin to change my clothes, the rain now over, I had to shut the door in her face to keep her out; when I opened it there she was; and she followed me to the windlass when I went to set the anchor-engine going: for I intended, I suppose, to take her to Imbros, where she might live in one of the broken-down houses of the village; but when the anchor was half up, I stopped the engine, and let the chain run again: for I said "No, I will be alone, I am not a child."

I knew that she was hungry by the look in her eyes: but I

He doesn't want the woman with him

cared nothing for that. I was hungry, too: that was all I cared about.

I would not let her be there with me another moment—got down into the boat, and, when she followed, rowed her back all the way past Foundoucli and the Tophana quay to where one turns into the Golden Horn by St. Sophia, round the mouth of the Horn being now a vast semicircle of charred wreckage, carried out by the river-currents; then, in the Horn, I went up the steps on the Galata side before one comes to where the barge-bridge had been; and when she had come after me on to the embankment, I passed up one of those mounting streets, encumbered now with stone-*débris* and ashes, but still marked by some standing wall-fragments, it being now not far from night, but the air as bright and washed with the rain and after-light of the sun as the blush of some purplish diamond, the west a Tyre aburn; and when I was two hundred yards up in this mixed quarter of Greeks, Turks, Jews, Italians, Albanians, and noise and cafedjis and wine-bibbing, I, having now turned two corners, suddenly gathered my skirts, spun round, and, as fast as I could, was off at a heavy trot back to the quay.

She was after me; but, being taken by surprise, I suppose, was distanced a little at first, though by the time I could scurry myself down into the boat, she was so close upon me, that she only rescued herself from falling into the water by balancing in her stoppage at the embankment-brink, as I pushed off.

I then set out to get back to the ship, muttering: "You can have Turkey, and I will keep the rest of the world," rowing seaweed with my face steadily averted from her, for I would not look to see what she was doing; but, as I turned the point of the quay where the open sea washes rough and loud, to row northward and vanish from her, I heard a babbling outcry—the first sound which she had uttered. I did look then: and she was still near me, for the silly maniac had been racing along the embankment, following me.

"Well, you little fool," I cried out across the water, "what are you after now?" and, oh, my God, shall I ever forget that strangeness, that wild strangeness, of my voice addressing under the sun another soul?

There she stood, whimpering like a dog after me: so I turned the boat round, rowed to the first steps, landed, and struck her two stinging slaps, one on each cheek.

While she cowered, surprised no doubt, I took her by the hand, led her back to the boat, rowed over to the Stamboul side, landed, and set off, still holding her hand, my object being to find some sort of house near by, not hopelessly eaten out by fire, in which to leave her: for in all Galata there was clearly none, and Pera, I thought, was too far to walk to. But it would have been better if I had gone to Pera, for we had to walk quite three miles from Seraglio Point all along woman injured trying to follow AJ

the city battlements to the Seven-towers, she picking her bare-footed way after me through the Sahara of charred stuff, and night now well arrived, the moon at large in the vast of heaven, rendering the lonesomeness of the ruins tenfold desolate, so that my bosom smote me then with bitterness, and I had a vision of myself that night which I will not write on paper. At last, however, pretty late in the evening, I got to see a mansion with a façade of green lattice-work, and a shaknisier, and terrace-roof, which had been hidden from me by the arcades of a bazaar—this bazaar being a vast space at about the center of Stamboul, one of the largest of the bazaars, I should think—in the middle of which stood the mansion, the home of some pacha or vizier, for it had a very distinguished look in that place; and it seemed little injured, though the vegetation which had choked the bazaar was singed to black fluff, among which lay thousands of calcined bones of man, mule, camel, horse: for all was illumined in that lucid, yet so pensive and forlorn, moonlight, that Orient moonlight of mystery that illumines Persepolis, and Babylon, and ruined cities of the Anakim.

The house, I knew, would contain divans, *yatags*, cushions, foods, and a hundred luxuries still good, for it was all shut in by a wall, though the foliage over the wall had been singed away, and the gate, all charred, gave way at a push from my palm; and now I crossed a court to the house, threw open a little lattice-door in the façade under the shaknisier, and entered. Here it was dark: and the instant that she, too, was within, out I slipped quick, slammed the door in her face, and hooked it upon her by a little hook over the latch.

I now walked out some yards beyond the court, then stopped in the bazaar, hearkening for her cry: but all was still; five minutes—ten—I waited: no sound; so now I continued my grum and melancholy way, hollow with hunger, intending to be off that night for Imbros.

But I had hardly advanced twenty steps, when I was aware of a strangled cry, apparently in mid-air behind me, and glancing back, beheld her through the gateway lying a white thing in black stubble-ashes, she having apparently jumped from a casement of lattice on a level with the little shaknisier-grating, through which once peeped bright eyes, twenty-five feet high.

I don't suppose that she was conscious of danger in jumping: for the laws of nature are new to her; and, having sought and found the opening, she may have just come naïvely after me as the cascade leaps and does not care. When I paced back and pulled at her arm, I found that she could not stand, her face screwed in mute pain, no moaning, her left foot bloody: and by the wounded foot I took her, and drew her so through the cinders of the court, and hurled her like a little cur with my whole force within the doorway, cursing her.

Now I would not trudge back to the ship, but struck a match, and went lighting up girandoles, cressets, candelabra, into a confusion of lights among a multitude of pale-tinted pillars, rose and azure, with verd-antique, olive, and Portoro marble, and serpentine; the mansion large: I having to traverse a desert of brocade-hangings, slim pillars, Broussa silks, before I spied a doorway behind a Smyrna *portière* at a staircase-foot, went up, and roamed some time about the house—windows with gilt grills, little furniture, but palatial spaces, hermit pieces of faïence, huge, antique, and arms, my footfalls muted in the Persian carpeting; till I passed along a gallery having only one window-grating that overlooked an inner court, and by this gallery entered the harem, which declared itself by a headier luxury, bric-â-bracerie, and baroqueness of manner; from which, descending a little stair behind a *portière*, I came into a species of larder paved with marble, in which grinned a negress in an indigo garb, her hair still adhering, and here an infinite supply of sweetmeats, French preserved-foods, sherbets, wines, and so on: so I put a number of things into a pannier, passed up again, found in the cavity of a garnet some of those pale cigarettes which drunken, then a jewelled chibouque two yards long, and tembaki; with all I descended by another stair, deposited them upon the steps of a kiosk of olive-marble in a corner of the court, passed up again, and brought down a *yatag* to recline on: and there by the kiosk-steps I ate and passed the night, smoking for hours in a state of lassitude, eyeing where, at the court's center, the alabaster of a square well blinks out white through a rankness of wild vine, weeds, acacias in flower, jasmines, roses, which overgrow both it and the kiosk and the whole court, raging too over the four-square arcade of Moorish arches round the court, under one of which I had hung a lantern of crimson silk; and near two in the morning I dropped to sleep, a deeper peace of gloom now brooding where so long the hobgoblin Mogul of the moon had governed.

* * *

When it was day I rose and made my way to the front, intending that that should be my final night in this place: for through the night, sleeping and waking, the thing which had taken place filled my brain, deepening from one depth of incredibility to a deeper, so that finally I arrived at a kind of conviction that it could be nothing but a drunken dream; but, as I opened my eyes afresh, the realization of that event flashed like a pang of lightning through my frame, and saying "I will go again to the far Orient, and forget," I set out from the court, not knowing what had become of her during the night; till, having arrived at the outer apartment, with a start I saw her lying there by the door, asleep sideways, head on arm, in the same spot where I had tossed her: so softly, softly,

I stept over her, got out, was off at a clandestine trot—the morning all in *fête*, very fresh and pure—and, after running two hundred yards to one of the bazaar-arches, I stopped, looking back to see if I was followed; but all that space was desolately empty, and I then walked on past the arch of ogives, the panorama of destruction now outspread before me—a few walls still standing, their windows framing the sky beyond, here and there a pillar or half-minaret, still some trunks without branches down within the Seraglio-walls, in Eyoub and Phanar branchless forests, on the northern horizon Pera still there; and, all between, blackness, stones, a rolling landscape of ravin, like the hilly pack-ice of the Pole, if its snow were ink; and to the right Scutari, black, laid low, with its suburb of tombs and some stumps of its woods, the sea brisk, blue-eyed, with its mob of *débris*-scum floating brown before the mouth of the Golden Horn: for I stood pretty up-lifted at the middle of Stamboul, somewhere in the region of the Suleimanieh, or of Sultan-Selim, as I judged, with vistas into abstract distances and mirage: but to me it looked too vast, too lonesome; and after advancing a score of yards be-yonds the bazaar, I yearned and turned back.

* * *

I found the creature still asleep at the house-door, and, kick-ing her, woke her: on which she sprang up with a start of sur-prise and quite a sinuous agility, to stare there at me, till, separating reality from dream and habit, she realized me, and then immediately fell afresh, in pain: so I hauled her up, and made her limp after me through several halls to the inner court and the well, where I set her among the bush on the alabaster, took her foot in my lap, examined it, drew water, washed it, and bandaged it with a rag torn from my caftan-hem, now and again talking gruffly to her, so that she might no more follow me.

After which I had breakfast by the kiosk-steps, and, when I had finished, put a mass of truffled *foie gras* on a plate, brushed through the thicket to the well, and gave it her. She took it, but looked foolish, not eating, so with my forefinger I put some into her mouth, whereupon she fell to devouring it; I also gave her some ginger-bread, a handful of bonbons, some Krishnu wine, and some anisette.

I then started out afresh, harshly telling her to stay there, and left her seated on the well, her hair hanging down the opening, she peering after me through the bush; but I had not half got to the bazaar-portal, when, glancing anxiously back, I saw that she was limping after me: so that this creature tracks me in the manner of a shell led away in the wake of a ship.

I now returned with her to the house, for it was necessary that I should excogitate some further method of dodging her.

Dressing the woman

That was four days ago, and here I have stayed: for the house and court are sufficiently agreeable, and are a museum of *objets d'art*. It is settled, however, that tomorrow I be off to Imbros.

* * *

It seems that she never wore, nor knew of, clothes, and it was only here and there that one could descry her ivory-brown color, the rest being crusted with dust like bottles long cobwebbed in cellars.

So I have dressed her, first sousing her with sponge and soap in lukewarm rose-water in the silver cistern of the harem-bath, a marbled apartment with a fountain and the intricate ceilings of these houses, and frescoes, and gilt texts of the Koran glinting on the marble and on the hangings of rose-silk. I had flung some clothes on a couch, and, having shewn her how to towel herself, made her step into a pair of the trousers called *shintiyan*, made of white-silk with yellow stripes, which, by a running string, I tied above the hills of her hips, then drawing up the bottoms to her knees, tied them there, so that their voluminous folds, overhanging to the ankles, have the look of a skirt; over this I put upon her a chemise, or quamis, of chiffon, reaching to the hips; then a jacket or vest of scarlet satin, embroidered in gold and precious stones, reaching to the waist, tight-fitting; and, making her lie on the couch, I slipped upon her little feet little baboosh-slippers, blue, then anklets, on her fingers rings, round her neck a necklace of sequins, finally dyeing her nails, which I cut, with henna; there remained her head, but with this I would have nothing to do, only pointing to the tarboosh which I had brought, to a kerchief, to some corals, and to the fresco of a woman on the wall, which, if she chose, she might copy; lastly, I pierced her ears with the silver needles which they used here, and after two hours of it left her.

An hour later I saw her in the arcade round the court, and, to my astonishment, she had a plait down her back, and round her brows a feredjeh, or hood, of sky-blue silk, precisely as in the picture.

* * *

Here is a question the answer to which would be interesting to me: Whether or not for twenty years—or say rather twenty centuries—I have been stark mad, a raving maniac; and whether or not I am now suddenly sane, seated here writing in my right mind, my whole tone changed or speedily changing? And whether such change may be owing to the presence of only one other being on the sphere with me?

* * *

This singular being! Where she has lived, and how, is a

problem beyond all solving. She had, I say, never seen clothes: for when I set to dress her her perplexity was endless; moreover, during her twenty years she has never seen almonds, figs, nuts, liqueurs, chocolate, conserves, vegetables, sugar, oil, honey, sweetmeats, orange-sherbet, mastic, salt, raki, tobacco, for she has shown perplexity at all these: but she has known and tasted *white wine*: I could see that. Here, then, is a mystery.

* * *

I have not gone to Imbros, but remained here some days longer, taking stock of her.

I have permitted her to sit in a corner of the apartment at meal-times, not far from where I eat; and I have given her to eat.

She is wonderfully clever! I continually find that, after an incredibly brief time, she has adapted herself to this or that, already wearing her clothes with a certain coquetry, as though a clothes-wearer by birth; and, without in the least *seeming* observant—for she gives an impression of giddiness—she reckons me up, I am convinced, closely: knows when I am talking roughly, bidding her go, bidding her come, sick of her, tolerant of her, scorning her, cursing her; nay, if I even wish her to the devil, she sees, and will disappear. Yesterday I noticed something queer about her, and discovered that she had been staining her lids with kohol, like the *hanums*: so that, having found some, she must have guessed its use from the pictures: wonderfully clever! imitative as a mirror. Again, two forenoons ago, on seeing a kittur of mother-of-pearl, I played an air, sitting under the arcade; I could see her, meanwhile, behind one of the pillars on the opposite side of the court listening closely, and, I fancied, panting; and, on returning from a walk beyond the Phanar walls in the afternoon, I heard the same air coming out from the house, she repeating it faultlessly by ear; also, during the forenoon of the day before, I came upon her—for footsteps make no sound in this house—in the pacha's visitors'-hall and what was she at?—copying the postures of three dancing-girls frescoed there! so that she would seem to have a character as flighty as a butterfly's, and troubles about nothing.

* * *

Now I know.

I had noticed that at the beginning of each meal she seemed to have something on her mind, going toward the door, hesitating as if to see whether I would follow, and then returning; and at length yesterday, after sitting to eat, she jumped up, uttering to my infinite surprise her first word—with a very experimental effort of the tongue, like a fledgling which tries the air: the word "*Come.*"

That forenoon, on meeting her in the court, I had told her to repeat some words after me; but she had made no attempt, as if shy to break the silence of her life; and now I felt some species of childish pleasure in hearing her utter that word, frequently no doubt heard from me: so, after hurriedly eating, I went with her, saying to myself "She must be about to shew me the food to which she is accustomed, and that may solve her origin."

And so it has proved. I have now discovered that, to the moment when she saw me, she had tasted only her mother's milk, dates, and that white wine of Ismidt which the Koran permits.

As it was getting dark, I lit and took with me the red-silk lantern, and we set out, she leading, walking confoundedly fast, slackening when I swore at her, then getting fast again: and she walks with a kind of levity, flightiness, liberated *furore*, very difficult to describe, as though space were a luxury to be revelled in. By what instinctive cleverness or vigor of memory she found her way I cannot tell; but she led me such a walk that night, miles, miles, till I became furious, darkness having soon fallen, with only a faint moon obscured by cloud, and a drizzle which haunted the air, she without light climbing and picking her thinly-slippered steps over piles of stone with a flying ply of foot, I anon dipping a foot with horror into one of those little ponds which always spotted the Stamboul streets. In the moments when I was nearer her I would see her peer upward toward Pera, as if that were a known landmark, would note the constant aspen caprices of the coral drops rocking in her ears, the nimble of her limbs, and would wonder with a groan if Pera was our goal.

Our goal was even beyond Pera. When we had got to the Golden Horn, she pointed to my caïque which lay at Old Seraglio Steps, and over the water we went, she lying quite at ease now, her face at the level of the water in the center of the caïque's crescent-shape, as nonchalant as a *hanum* of old, engaged in some escapade, going over to the Babel of Galata and that north bank of the Horn.

Then through Galata we passed, I already cursing the journey; and, following the line of the coast and that steep thoroughfare of Pera, we came at last, almost in the country, to a great wall, and to the entrance to a great terraced garden, whose limits were invisible, many of the avenues being still intact.

I knew it at once—had laid a special fuse-train in the palace at the top of the terraces: the royal palace, Yildiz.

Up and up we mounted through the grounds, a few unburned persons in rags of uniform still discernible at random, as the lantern swung past them: a musician in blue, a fantassin in scarlet, three domestics of the palace in red-and-orange . . .

The palace itself was all a ruin, together with all its surrounding barracks, mosque, seraglio, and, when we got to the top of the grounds, presented a picture very like those I have seen of the ruins of Persepolis, only that here the columns, both standing and fallen, were innumerable, and all more or less blackened; and through doorless doorways we moved, down flights of four or five steps immensely-wide, and up them, and over strewn courtyards, by tottery fragments of arcades, all roofless, and tracts of charcoal reaching away between the relics of avenues of columns, I following, expectant, her feet very keen now. Finally, down a flight of narrow steps, very dislocated, we jolted to a level which, I thought, must be the floor of the palace vaults: for at the foot of the steps we stood on a plain of plaster, which shewed the marks of the flames; and over this the girl spurted, pointing with eager recognition to a hole in it, and disappeared down the hole.

When I, on following to the hole, lowered the lantern into it, I saw that the drop down was about eight feet, made less than six feet by a heap of stone-rubbish below, the falling of which had caused the hole: and it was by standing on this rubbish-heap, I knew at once, that she had managed to climb out under the sky.

Dropping down now, I found myself in a cellar with a floor of marl, fusty and damp, but so very vast in area, that even in the day-time, I believe, I could not have made out its limits: for I think that it extends under the whole palace and its environs—a stretch of space of which, with the lanterns, I could only see a little portion.

She still leading me keenly on, I presently came upon a region of boxes, each about two feet square, nine inches high, made of flimsy laths, packed to the roof; and two hundred feet from these I saw, where she pointed, a region of bottles, bottles with paunches in chemises of wicker-work, stretching away into dimness and invisibility: the boxes, of which a throng lay broken open, as they can be by just wrenching at a crack, containing dates, and the bottles, of which many thousands lay empty, containing old Ismidt wine. Some fifty or sixty casks—covered with mildew—some broken bits of furniture, a cube of parchments—large as a cottage, rotting, curling—showed that this cellar had been more or less loosely used for the storage of unwanted odds-and-ends.

It had been used, too, as a domestic prison: for in the lane betwixt the region of boxes and the region of bottles there lay the skeleton of a woman, the details of whose costume were still appreciable, she having thin shackles of brass on her wrists: and when I had scrutinized her I knew the history of the being standing silent by my side.

This being is a daughter of the Sultan, as I assumed when I had once understood that the skeleton is both the skeleton of her mother, and the skeleton of the Sultana.

That the skeleton was her mother is evident: for when the

cloud came, twenty years since, the woman was in the prison, which must have been air-tight, and with her the girl; and since the girl is certainly not over twenty—she looks younger—she must have been either unborn or a baby: but a baby would hardly be imprisoned with another than its mother. I rather think that the girl was unborn at the moment of the cloud, and was born in the cellar.

That the mother was the Sultana is evident from her fragments of dress, and the symbolic character of her every ornament—crescent ear-rings, heron-feather, and the blue campaca enameled in a bracelet—this poor woman having perhaps been the victim of some fit of imperial spleen, envenomed by some domestic misdemeanor which may have been pardoned in a day, had not death overtaken her master and humanity.

There are five steps near the center of the cellar, leading up to a trap-door of iron, at present fastened, this apparently being the only opening into this hole: and this trap-door must have been so almost air-tight as to exclude the intrusion of the poison in deadly quantity.

But how rare—how strange—the coincidence of chances here. For, if the trap-door was quite air-tight, I cannot think that the supply of oxygen in the cellar, large as it was, would have been sufficient to last the creature twenty years, to say nothing of what her mother breathed before death: for I assume that the woman must have continued to live some time in her dungeon, sufficiently long, at least, to teach her child to acquire its fare of dates and wine: so that the door must have been only just hermetic enough to block the poison, yet admit some oxygen—unless the place was quite air-tight at the time of the catastrophe, and some crack which I have not observed, due perhaps to earthquake, opened to admit oxygen and some sunlight after the poison was dissipated: in any case—the all-but-infinite rarity of the probability!

Thinking these things I climbed out, and we walked to Pera, where I slept in a white-stone house in five or six acres of garden overlooking the cemetery of Kassim, having pointed out to the creature another house in which to sleep.

This creature! what a history! After existing twenty years in a sunless universe hardly nine acres wide, she one day saw the only sky which she knew collapse at one point! a hole opens into yet a universe beyond! It was *I* who had arrived, and fired a city, and set her free.

* * *

Ah, I see something now! I see! it was for this that I was preserved: I to be a species of First-man, and this creature to be my Eve! That is it! *"The White"* does not admit defeat—would recommence the race—at the final, the eleventh hour, in spite of everything, would twist rout into victory and outwit that Other.

AJ will not act to revive the race of mankind

However, if this be so—and I seem to see it—then, in that White scheme is a flaw; at *one* point that elaborate Forethought rambles: for I am such, that I choose to refuse.

Certainly, in this matter I am on the side of "Black": and since it depends absolutely upon me, this time Black wins.

No more men down this way after me, ye Powers! To *you* the question may be nothing more than a gaming-table exhilaration as to the outcome of your aërial squabble, but to the poor beggars who had to bear the racks, rack-rents, wrongs, sorrows, horrors, it was strong stuff, you know! Oh, the deep, deep pain—the commonness and dullness—of that bungling ant-hill, now happily wiped out! My darling Clodagh—not ideal! Those lubber "lords" and "ladies" of my day! And there was a man named Judas who "betrayed" that gentle Jesus, and some Roman dog named Galba, and a French devil, Gilles de Raiz: and the rest were much the same. No, not a good race, that small infantry that called itself Man; and here, falling on my knees before God and Devil, I vow: Never through me shall it sprout and fester afresh.

* * *

I cannot realize her! Not at all, at all, at all! If she is out of my sight five minutes, I fall to doubting her realness; if I lose her during two hours, all the old feelings, like certainties, recur, that I have merely been dreaming—that this appearance cannot be an objective fact of experience, since the impossible is impossible.

Seventeen years, long years of madness. . . .

* * *

Tomorrow I start for Imbros: and whether this being chooses to follow me, or whether she stays here, I will see her from the moment I am there no more.

* * *
* * *

She must rise very early. I who am now regularly on the palace-roof at daybreak, from the silks of the galleries, or from the steps of the telescope-kiosk, may detect her away down below, a microscopic form running about the sward, or staring up in wonder at the palace from the lake's border.

When three months ago she came with me to Imbros, I left her in that house in the village with the green jalousies facing the beach, where there was everything that she would need; but I knew that, like all the houses down there now, it leaked profusely: so the next day I went down to that stair cut through the cliff-rock south of the village, climbed it, and half a mile onward found a park and villa which I had seen from the sea, the villa almost intact, strongly built of porphyry, though small, and very like a Western house, with shingles, and three gables, so that I think it may have been the yali of some Englishman, for it has English books, though the only person I observed there was an Aararat Kurd, with ankle-pantaloons and shoulder-cloak; and all in the park, and all about the rock-steps, growths of mandragora, and from the rock-steps to the house an avenue of acacias, mossy underfoot, that join in an arch overhead, the house standing about four yards from the brink of the sea-cliff, whence one can see the *Speranza's* main-topmast in her haven. Then after examining the place I went down again to the village and her house; but she was not there; and two hours long I paced about among the bush of these amateur little alleys and flat-roofed houses without windows (though some have terrace-roofs and a rare aperture), whose yellows, reds, blues, once crude, look now like sunset hues when the flush has just faded, and they faint away. When at last she cafe running with her lips split, I took her up the rock-steps to the villa; and there she has lived, one of its roof-tips, I now find, being just visible from the northeast corner of the palace-roof, two miles from it.

That evening afresh, when I was leaving her, she made an attempt to follow me; but I was resolved to end it then: so, plucking a sassafras-whip, I cut her deep, three times, until she ran crying.

* * *

So, then, what is my fate henceforth?—to think always, from sun to moon, and from moon to sun, of one only thing, and that thing a mite for the miscroscope? to evolve into a Paul Pry to spy upon the hoppings of one sparrow, like some fatuous gossip of old, his greed to peep, his sole faculty to sniff, his glee and his victory to unearth the infinitely insignificant? I would kill her first!

* * *

I am convinced that she is no stay-at-home, but roams continually over the island: for thrice, roaming myself, I have lighted upon her, she that first time rushing with a flushed face, bent upon striking down a butterfly with a bush held in the left hand (for both hands she uses with dexterity)—about ten in the forenoon it was, in her park, at the lower end where grasses grow rank, and there is a hypertrophy of fernery luxuriating between the tree-trunks, and obscurity, and the broken wall of a funeral-kiosk sunk askew under moss, creepers, and wild flowers, behind which I peeped concealed, soaked with dew. She has had the assurance to modify the dress I put upon her, and was herself a butterfly: for, instead of the shintiyan, she had on baggy pantaloons of azure silk, a zouave of saffron satin hardly reaching to the waist, no feredjé, but a fez with violet tassel, her plait quite tidy, but her forehead-hair wanton, the fez cocked backward, while I got glimpses of her careening heels lifting out of the dropping slipper-sole and she is pretty clever, but not clever enough, for that butterfly escaped, and in one instant I saw her alter into weary and triste, for in Nature is nothing more fickle than that face, which is like a landscape swept with cloud-shadows on a sunny day. Fast beat my heart that morning, owing to my consciousness that, while I saw, I was unseen, yet might be seen.

And three weeks afterwards I came upon her at noon a good way up yonder, west of the palace, asleep on her arm in an alley between trellises, where rioting wild-vine that overgrew them buried her in gloom; but I had not been peering through the bush three minutes, when up she starts to look ardently about, her quick consciousness, I suspect, having detected a presence, though I think that I contrived to win away unseen. I saw that she keeps her face pretty dirty, all about her mouth being dry-stained with a polychrome of grape, *mûrs*, and other colored juices, like slobbering *gamins* of old; I could also see that her nose and face are at present sprinkled with little freckles.

Five evenings since, on seeing her a third time, I observed that the primitive instinct to represent the world in *pictures* has been working within her: for she was drawing. It was down in the village, whither I had strolled, and on coming out upon a street from an alley, saw her near, pulled up short, and peered at her on her face all among bush, a bit of board before her, in her fingers a chalk-splinter, and intently she was drawing, her tongue-tip travelling along her short upper-lip from side to side, regularly as a pendulum, her fez tipped far back, her left calf swinging upward from the knee. She had drawn her yali, and now, as I could see by peering far forward, was drawing the palace from memory, for there were the waving lines meant for the platform-steps, the two pillars, the battlements of the outer court, and before the portal—my turban reaching above the roof, my two sheaves of beard sweeping below my knees—myself.

Something pricked me, and I could not resist uttering a "Hi!", whereupon she scrambled like a chamois upright, I pointing to the drawing, smiling.

This being has a way of pressing her lips mincingly, while she shakes her face at me, cooing a fond sort of laugh—as she cooed now.

And I: "You are a clever little wretch, you know"—she cocking her right eye, trying to divine my mind with a kind of smile.

"Yes, a clever little wretch," I went on in a rough voice, "clever as a serpent, no doubt: for in the first case it was the Black who used the serpent, and now it is the White: but it will not work this time. Do you know what you are to me, you? My Eve!—a little fool, a little piebald frog like you.

But it will not do at all! A nice race it would be with you for mother, and me for father, wouldn't it?—half-criminal like the father, half-idiot like the mother: like the last, in short. They used to say, in fact, that the offspring of a brother and sister was always weak-headed and from such a wedlock came our race, so no wonder it was what it was: and so it would have to be again now. Well, no, whatever cares we take, the White will trick us: so no risks—unless we have the children, and cut their throats at birth; but *you* would not like that at all, I know; and, on the whole, it would not work, for the White would be striking a poor man blind with His lightning, if I tried that on. No, then: the modern Adam is some six hundred thousand years wiser than the first—you see? less instinctive, more rational. The first 'disobeyed' by commission; I by omission; only his 'disobedience' was a 'sin', mine is a heroism. I have not been a particularly ideal species of beast so far, you know: but in me, Adam Jeffson—I swear it—the race shall at last attain to nobility, the nobility of self-extinction. I shall turn out trumps; shall prove myself stronger than Tendency, World-Genius, Providence, Currents of Fate, White Power, Black Power, or whatever is the name of it. No more Clodaghs, Borgias, 'lords', Napoleons, Peaces, Rockefellers, Hundred-Years' Wars—you see?"

She kept her eye obliquely cocked upward like a little fool, wondering, no doubt, what I was saying.

"And, talking of Clodagh," I went on, "I shall call you that henceforth, to keep me reminded. So that is your name—not Eve—but Clodagh, who was a Poisoner, you see? She poisoned a poor man who trusted her: and that is your name now—not Eve, but Clodagh—to remind me, you most perilous little speckled viper! And in order that I may no more see your foolish little pretty phiz, I decree that, for the future, you wear a *yashmak* to cover up your lips, which, I can see, were meant to be seductive, though dirty; and you can leave the blue-blue eyes, and the little nose, with the freckles on its white skin, uncovered, if you like, they being commonplace enough. Meantime, if you care to see how to draw a palace—I will show you."

Before I stretched my hand she was presenting the board—so that she had perceived something of my meaning! but somewhat of guttural in my tone had wounded her, for she presented it looking glum, her under-lip pushing crooked out, very pathetically, I must say, as usual when she is inclined to cry.

Well, in a few strokes I drew the palace, and herself standing at the portal betwixt the pillars: and now great was her satisfaction, for when she pointed to the figure and then to herself interrogatively, and I nodded "yes," she went cooing her fond monotone with closed lips mincing; and it is clear that, in spite of my beatings, she but slightly fears me.

Before I could move away I felt some rain-drops, and down

in some seconds rushed a shower; also I saw that the vault was fast darkening, so I darted into the nearest of the piggeries, leaving her glancing sideways skyward with the qaintest interest in the rain: for she is not yet familiarized with things, and seems to regard them with an artless seriousness and curiosity, as though they were living things, comrades as good as herself. Even when she presently joined me, she reached out to feel the drops.

Now there tumbled out a thunder-clap, a wind was blowing up, rain spraying about me: for these wee box-houses' window-panes (made, I believe, of paper saturated in almond-oil), have long disappeared, and rains, penetrating by roof and rare window, splash the bones of men: so I was gathering up my skirts to rush toward other shelter, when she spurted from the door to me, saying in her experimental utterance that word of hers: "*Come,*" and ran out in advance, while I, tossing my external robe over my turban, followed, to urge my way against the scourge of the rain-wash.

She took the way, by the horse-pond, through an alley between two walls, then down a path through wood to the rock-steps; and up we ran, and along the hill, to her yali, which is a mile nearer the village than the palace is, though by the time we pelted into its shelter we were wet to the skin.

Sudden darkness had come; but she quickly unearthed some matches, lit one, looking at it with a certain air of meditation; then applied it to a candle and to a bronze Western lamp on the table, which I had taught her to oil and light; and when I pointed to a mangal like one which she had seen me light to warm bath-waters in Stamboul, she ran to the kitchen, ran back with some sticks, and very clearly lit them. And there for hours I sat that night reading (the first time for many a year): reading a book by the poet Milton, found in a book-case on the other side of the Western fire-place by which the mangal stood: and most strange, most novel, I found that oratory about Black Power and White Power and warring angels that night, while the storm raged: for this man, though scant in brain-power like the ancients in general, had evidently taken no end of pains with his book, and done it wonderfully well, too, making the thing hum; and I could not conceive why he should have been at that trouble, unless it was for the reason that I reared the palace—some spark in a man—and he would be like the Gods—but that is vanity.

Well, there is a venom about the tempests recently that really transcends bounds; I believe I have noted it in these sheets before—I never could have convinced of turbulences so huge, such as I heard them that midnight seated there smoking a chibouque, reading, listening to the bawlings and lamentations of that haunted air, shrinking from it, fearing even for the *Speranza* by her quay in the harbor, and for the palace-pillars. But what astonished me was that female thing: for after being seated on the ottoman to my right some time, she dropped sideways asleep, not the least fear about her, though I should have thought that nervousness at such a turmoil would be so certain to occur in her; and whence she has this nonchalant confidence in the cosmos into which she has so suddenly come I do not know: for it is as though someone inspired her with the mood of lightness, saying "Be of good cheer, and care not a fig about anything: for God is God."

I heard the ocean hawking hoarse, hurtling like heavy ordnance against the bluffs below, where the seas meet the southern of the two claws of land that form the harbor; and the thought came into my head: "If, now, I taught her to speak, to read, I could sometimes make her read to me."

The winds were wilfully wrestling with the villa to wring it away into the drear infinities of the night, and I could not but heave a sigh: "Alas for us two cast-aways of our race, pieces of flotsam tossed up here a moment, ah me, on this coast of the æons, soon to be hauled back, O eternity, down the Bottomless of you turbid maw; and upon what strand—who shall say?—shall she next be tossed, and I, separated then perhaps by the stretch of the astral tract?"; and such a pity, and a wringing of the heart, seemed in things, that a tear parted from me that dismal midnight.

She started up at a wrath of more appalling volume, rubbing her eyes, with untidied hair (it must have been about midnight), listening a minute with that demure droll interest of hers to the turmoil; then smiled to me; rose then and left the room, presently to come again with a pomegranate and some almonds on a plate, some delicious rich liquor, too, in an Ægean cruche, and a silver cup, gilt inside, standing in a zarf; these she placed on the table at my hand, I murmuring "Hospitality."

And now she stood looking at the book, which I read as I ate, with her left eye-lid lowered, trying to divine its use, I suppose. Most things she understands quick, but this must have baffled her: for to see one looking fixedly at a thing, and not know what one is looking at it for, must be very disconcerting.

So I held it up before her, saying: "Shall I teach you to read it? If I did, how would you repay me, you Clodagh?"

Upon which she cocks her eye, trying to comprehend, the candle-flame, moved by the wind like a brush which paints, flickering on her face, though every cranny was closed; and, God knows, at that moment I pitied the dumb waif, alone in the whole globe with me.

"Perhaps, then," I said, "I will teach you. You are a pitiable little derelict of your race, you know; and two hours every day I will let you come to the palace, and will teach you. But be sure, be careful, if there be danger, I will kill you—assuredly

—without fail; and let me begin with a lesson now: say after me: 'White.' "

I took her hand, got her to understand that I wanted her to repeat.

"White," said I .

"Hwhite," says she.

"Power," said I.

"Pow-wer," said she.

"White Power," said I.

"Hwhite Pow-wer," said she.

"White Power shall not," said I.

"Hwhite Pow-wer sall not," said she.

"Prevail," said I.

"Fffail," said she.

"Pre-vail!" said I.

"Pe-vvvail," said she.

"White Power shall not prevail," said I.

"Hwhite Pow-wer sall not—fffail," said she.

A thunder which roared as she uttered it seemed to me to go guffawing through the cosmos, and a minute I gazed upon her face with positive fear; till, starting up, I thrust her from my path, and darted forth to battle my way to the palace and my bed.

Such was the ingratitude and fatality which my first attempt, five nights since, to teach her met with; and now it remains to be seen whether my pity for her dumbness, or some servile tendency toward fellowship in myself, will result in any further lesson. Certainly, I think not: for though I have given my word . . . we shall see.

Surely her presence in the world with me—for no doubt it is that—has worked some profound modifications in my mood: for gone now are those storm-tossed hours when, stalking like a cock, I flaunted my monarchy in the face of the heavens with blasphemies, or else dribbled, shaking up my body in a lewd dance, or was off to reduce some city to ashes and revel in redness and the chucklings of Hell, or rolled in the drunkenness of drugs. It was frenzy!—I see now—it was "not good," "not good." And it rather looks as if it were past —or passing. I have clipped my beard and hair, taken out the ear-rings, and thought of modifying my raiment . . . I will watch to see whether she comes loitering down there round the gate of the lake.

* * *

Her progress is like . . .

* * *

* * *

It is some nine months since I wrote that "Her progress is like . . . ," and have since had no impulse to write; but I was thinking just now of the tricks and eccentricies of my memory, and, seeing the old book, will record it here: for I have lately been attempting to recall the name of my old home in Britain, where I was born and grew up, and it is gone, gone; maybe it will come back to me later: for I can't say that my memory is bad; there are things—trivial little things sometimes—that come back to me with considerable vividness: for instance, I remember to have met in Paris (I think), long before the poison-cloud, a little Brazilian boy of the color of coffee-and-milk, whom she now constantly recalls to me: he wore his hair so close-cut, that one could spy the fish-white flesh betwixt, delighted to play by himself about the stairs of the hotel costumed in the spectral balloon-dress of a Pierrot, and I have the impression now that he must have had very large ears—clever as a flea he was, able to gabble six or seven languages, as it were by nature, without having any suspicion that that was at all extraordinary. She has the same light, unconscious, nonchalant cleverness, and easy way of life. It is little more than a year since I commenced to teach her, and already she can speak with a considerable vocabulary (though she does not pronounce the letter "r"); for chemistry she has a craving, a rage, and no little knowledge of it; she has also read, or rather devoured, many books; can write, draw, play the harp: and all she does without effort, rather with that flighty naturalness with which larks took to the wing.

What made me teach her to read was this: one afternoon, some fourteen months ago, I from the roof-kiosk saw her down at the lake-brink, a book in hand, and as she had beheld me looking steadily at books, so she was looking steadily at it, with her head held sideward, rather pathetic, so that I had to laugh: for I spied her through the glass; and whether she is the simplest little goose or the craftiest of rascals I am not yet sure. If I thought that she has the least design upon my honor, it would be ill for her.

I went to Gallipoli for three days in May, and came back bringing a pretty little caïque, a crescent of the color of the moon, which I fetched up in the motor to the lake after two days' labor in cutting a passage through bush-thicket; and it has pleased me to see her lie amid the silks at its middle, while I, plying the paddle a little, heard her say her first words—between eight and ten in the evening it was, though later it became 10 a.m. to noon when the reading began, we seated then on the palace-steps before the portal, her mouth covered with the yashmak, the lesson-book a Bible with large letters which I chanced to find at her yali. *Why* she must wear the yashmak she has never once asked; and how much she conjectures, knows, or intends, I have no notion, continually questioning myself as to whether she is all simplicity, or all depth.

That she is conscious of some profound contrast in our structure I cannot doubt: for that I have a long beard, and she none at all, is among the most obvious of facts.

* * *

I have wondered whether a certain *Western-ness*—a growing modernity of tone—may be the result, as far as I am concerned, of her presence with me? I do not know. . . .

* * *

There is the sheen of a lake just visible in the north forest from the palace-top, and in its fish like carp, tench, roach, &c., so in May I searched for a tackle-shop in the Gallipoli Fatmeh-bazaar, and got four 12-foot rods, with reels, silk-line, quill-floats, some silk-worm gut, with a packet of No. 7 hooks, and split-shot for sinkers; and, since red-worms, maggots, gentles are common on the island, I felt sure of more fish than I wanted, which was none at all: so, for the amusement, I fished several times, lying at my length in a patch of long-grass over-waved by an enormous cedar, where the bank is steep, and the water deep; and one afternoon she was suddenly there with me, questioned me with her eyes, and, when I consented, stayed: so presently I said I would teach her bottom-angling, and sent her heels flying up to the palace for a second rod and tackle.

But that day nothing was done: for, after teaching her to thread the worm and put the gentles on the hooks, I sent her to hunt for worms to chop up for ground-baiting the pitch for the next afternoon, and when this was done it was dinner-time: so I sent her home, for I was then giving the lessons in the morning.

The next day, however, I found her at the bank, taught her to take the sounding for adjusting the float, and she lay down not far from me, holding the rod. So I said to her: "Well, this is better than living in a cellar for years, with nothing to do but walk up and down, sleep, and consume dates and Ismidt wine."

"Yes!" says she.

"Year after year!" I said: "how did you bear it?"

"I was not closs," says she.

"Did you never suspect that there was a world outside that cellar?" I said.

"No," says she, "or, lather, yes: but I did not suppose that it was *this* world—another where he lived."

"He who?"

"You ask? He who told me—— Oh! a bite!"

I saw her float bob under, so, spurting to her, taught her how to strike and play it; and though it turned out to be only a

She tells of her experience in the cellar

tiny barbel, she was in ecstasies, stooping upon it on her palm, murmuring her fond coo.

Then, rebaiting, we lay again; and I said: "But what a life: no exit, no prospect, no hope——"

"Plenty of *hope!*" says she.

"Heavens! of what?"

"I knew vely well that something was lipening over the cellar, or under, or alound, and would come to pass at a certain fixed hour, and that I should see it, and feel it, and it would be vely nice."

"Well, you had to wait for it, anyway. Didn't those years seem *long?*"

"No—sometimes—not often. I was always occupied."

"In doing what?"

"Eating, dlinking, lunning, talking——"

"To your*self?*"

"Not to myself."

"To whom, then?"

"Why, to the one that told me when I was hungly, and placed the dates there."

"I see. . . . Don't wriggle about, or you will never catch any fish: the maxim of angling is 'Study to be quiet'——"

"O! another!" she called, and this time, all alone, very agilely landed a roach.

And presently I: "But do you mean that you were never sad?"

"Sometimes I would sit and cly," says she—"I did not know why. But if that was 'sadness,' I was never miselable, never, never. And if I clied, it did not last long, I would fall to sleep, for my love would lock me in his lap, and kiss me."

"Which *'love'?*"

"You ask that? But you know! He who told me when I was hungly, and of the thing that was lipening outside the cellar."

"Aha! I see. . . . But in that darkness—were you never afraid?"

"*I!* Of what?"

"Of the unknown."

"Now, how could I be aflaid? The known was the vely opposite of tellible: hunger and dates, thirst and wine, desire to lun and space to lun in, desire to sleep and dleams, yes, dleams! dleams! in sleep: the opposite of tellible; and the unknown was even less tellible than the known: for it was the nice thing that was lipening outside the cellar. How could I be——?"

"Ah, yes," said I, "you are a clever little being, no doubt, but your continual fluttering about is fatal to all angling. Isn't it in your nature to keep still a minute? And as to your habits in the cellar——?"

"Another!" she cried with a happy laugh, landing a young chub; and that afternoon caught seven to my one.

* * *

Another day I took her from the pitch to one of the kitchens in the village with some of the fish, until then always thrown away, and taught her cooking: for the only cooking-implement in the palace is the silver alcohol-lamp for coffee and chocolate; so we both scrubbed the utensils, and boil and fry I taught her, and the making of a sauce from vinegar, bottled olives, and American butter from the *Speranza*, and the boiling of rice mixed with flour for ground-baiting our pitch; upon which she, at first astonished, was presently all deft housewifeliness, breathless officiousness, and of her own instinctiveness grated some almonds lying there, with which to sprinkle the fried carp. We ate them sitting on the floor together: the first new food, I suppose, save fruits, tasted by me for twenty-one years; nor did I find it disagreeable.

The next day she came up to the palace reading a book which turned out to be a cookery-book in English, found at her yali; and a week later she appeared, out of hours, presenting me a dish of yellow delf containing a mess of gorgeous colors—a boiled chub buried beneath red of pepper, fragments of saffron, a greenish sauce, and almonds, but I sent her away, and would have none of her, or of her dish of fish.

* * *

Two miles up, west of the palace, is a ruin in forest, I think of a mosque, though only three pieces of pillars under creepers, and the weedy floor, with the courtyard and steps, remain, before it being an avenue of cedars, the path between the trees choked with long-grass and wild rye reaching to my middle; and here I saw one day a disc of brass, bossed in the middle, which may have been either a shield or part of an antique cymbal, with rings running round it from middle to circumference: so the next day I brought nails, a hammer, a saw, and a box of paints from the *Speranza*, painted the rings in different colors, cut down a lime-trunk, nailed the disc to it, and planted it before the steps: for I said I would make a bull's-eye, and do firing-practice down the avenue; and this the evening afterwards I was doing at four hundred feet, startling the island with that unusual alarum, when up she comes peering with inquiring eyes: at which I was cross, because my arm, long unused, was firing wide; but I was too proud to say anything, let her look, and soon she understood, laughing every time I made a considerable miss, until at last I turned upon her saying: "If you think it so easy, you may try."

She had been wanting to try, for she came spryly to the offer; and after I had opened and showed her the mechanism, the cartridges, and how to shoot, I put into her hands one of the

Speranza Colt's: upon which she took her bottom-lip between her teeth, shut her left eye, vaulted out the revolver to the level of her intense right eye, and sent a ball through the center of the boss.

However, it was a fluke-shot, for I had the satisfaction of seeing her miss every one of the other five, except the last, which hit the black. That, however, was three weeks since, and now my hitting record is forty per cent, and hers ninety-six—most extraordinary: so that it is clear that this creature is the *protégée* of someone, and favoritism is in the world.

* * *

Her book of books is the chemistry-book, and next the Old Testament. Sometimes, at noon or afternoon, I may look abroad from the roof or galleries, and see a remote figure seated on the sward beneath the shade of plane or cedar: and I can always divine that the book she cons there, away from her laboratory, is the Bible—like an old Rabbi: has a passion for stories, and there finds a store.

Three nights ago when it was already quite late, and the moon very glorious, I noted her moving homewards close to the lake, and howled down to her, intending to say "Good-night"; but she thought that I had called her, and came: and, sitting out on the stop step, we talked for hours, she without the yashmak.

And, talking about the Bible, says she: "What did Cain to Abel?"

"Knocked him over," I replied, liking to use such idioms, with the double object of teaching and teasing her.

"Over what?" says she.

"Over his heels," I said.

"I do not complehend!"

"He killed him, then."

"That I know. But how did Abel *feel*?"

"Oh, well," I said, "you see bones all round you: the same thing happened to them as to a fish when it lies all still."

"And the men and the fish feel the same after?"

"Precisely the same—lie in a stark trance, and dream a nonsense-dream."

"That is not dleadful. Why were men so aflaid?"

"Because they were all such cowards."

"Oh, not all! not all! far flom cowards."

(This girl, I know not with what motive, has now definitely set herself up against me as the defender of the dead race—with every chance she is at it).

"Many, anyway," I said: "tell me one who was not afraid——"

"Why, they fought in wars—for nothing," says she: "look at Isaac, when Ablaham laid him on the wood to kill him, he did not jump up and lun to hide."

She defends mankind against his pessimistic view

"Well, but," I said, "in books you read of the best people, but there were millions of others—especially about the time of the cloud—on a lower level—common, dull, lubberly, mean, debased, diseased, making the earth a murrain of vices and crimes."

This she did not immediately answer, seated with her back half-toward me, cracking almonds between her teeth, continually hitting one step with the ball of her stretched slipper, her fez and corals reflected as a blotch of florid red in the gold; then she bent aside and drank wine from the gold Javan goblet which I had brought from the temple of Boro Budor, her head covered by it; then, the little hairs at her lip-corners still wet, says she: "Vices and climes, climes and vices—always the same. But was that the point? The point was their cleverness—to find out what water is made of—to fly on those things—what a pletty, witty thing a ship is!—to find out that the atmosphere of Mars has more oxygen than ours—to talk acloss the continents—how inspired! If they were clever enough for all that, in time they would have been clever enough to find out how to live together. What were these climes and vices?"

"Robberies of a hundred sorts, murders of——"

"What made them *do* them?"

"Their lubber souls."

"But *you* are of them, *I* am, yet you and I live here together, and do no vices and climes."

Her astonishing shrewdness! "No," I said, "we lack *motive*. There is no danger that we should hate each other, for we have plenty of dates, wines, and thousands of things—our danger is rather the other way: but *they* hated because they were numerous, and there arose among them a question of dates and wine."

"Was there not enough land to glow dates and wine for all?"

"There was—yes. much more than enough; but some got hold of lots of it, and, as the rest felt the pinch of scarcity, there arose a pretty state of things—including the dulness and commonness, the vices and crimes."

"Ah, but then," says she, "it was not to their bad souls that the vices and climes were due, but to this question of land. If there had been no such question, there could have been no vices and climes, since you and I, who are just like them, do no vices and climes here, where there is no such question."

The limelight of her mind! Right into the heart of a matter does her wit drive quick.

"That may be so," I said; "but there *was* that question of land, as there always must be where millions with varying degrees of greed and luck and cunning live together."

"Oh, not necessalily!" she cried pressingly: "not at all, since there is more land than enough: for, if there should spling up more men now, and they, having the expelience of the past at their hand, made an allangement among themselves that the first who tlied to take more than he could work should be sent to dleam a nonsense-dleam, the question could never again alise!"

"It arose before—it would arise again."

"But no! I can guess how it alose before: the land was at first so vely, vely much more than enough for all, that the first men did not take the tlouble to make an allangement among themselves: and afterwards the habit of carelessness was confirmed; until at last the vely oliginal carelessness must have come to have the look of an allangement. But now, if more men would spling, they would be taught——"

"Ah, but no more men will *spling*, you see——!"

She was silent awhile; then: "There is no telling; I sometimes feel as if they must, and shall: the tlees bloom, the thunder lolls, the air makes me lun and leap, the glound is full of fluitfulness, and I hear the voice of the Lord God walking all among the tlees of the folests."

As she uttered this, I could see her under-lip push out askew and shiver, as when she is nigh to crying, and her eyes spring liquid; but in a moment more she looked at me full and smiled, so mobile is her countenance; and, as she looked, it suddenly struck me what a noble structure of a brow the creature owns, almost pointed at the uplifted summit, and broadening down bell-shaped, draped in strings of frizzy hair, which anon she shakes away with her head.

"Clodagh," I said after some minutes—"do you know why I called you Clodagh?"

"No? Tell me?"

"Because once I had a lover called Clodagh, and she was a . . ."

"But tell me first," she cried: "how did one know one's lover, one's wife, flom all the others? There were many faces —all alike——"

"Oh, there were little differences."

"Still, it must have been vely clever to tell: I can hardly fancy any face, except yours and mine."

"Because you are a little goose, you see."

"What was a goose like?"

"Thing like a butterfly, only bigger, and it kept its fingers spread out, with a skin between."

"Leally? How caplicious! And I am like that?—But what were you saying that your lover, Clodagh, was?"

"A Poisoner."

"Poisoner . . . And you call *me* Clodagh?"

"To remind me: lest you—lest you—should become my—lover, too."

"I *am* your lover."

"What, girl?"

"Do I not love you, who are mine?"

"Come, come, don't be a little—Clodagh was a *poisoner. . . .*"

"Why was she? Had she not enough dates and wine?"

"She had, yes: but she wanted more, more, the village bumpkin."

"So the vices and climes were not confined to those that lacked things, but were done by the others, too?"

"Aye."

"Then I see how it was!"

"How was it?"

"The others had got *spoiled*: the vices and climes must have commenced with those who lacked things, and then the others, always seeing vices and climes lound them, did them, too—as when one olive in a bottle is lotten the whole lump becomes collupted: and all thlough a little carelessness at the first; but if more men spling now——"

"But I *told* you, didn't I, that no more men will spring? You know, Clodagh, that the earth produced men by an eternal process, commencing with a low type of life, and cumulatively developing it, till at last a man stood up; but that can never occur again: for the earth is old, old, and has lost her evolving fervors now. So talk no more of men *splinging*, and of things which you do not understand. Instead, go inside—stay, I will tell you a secret: today in the wood I plucked some musk-roses and wound them into a wreath, meaning it for a crown for your forehead tomorrow, and it lies now on the pearl tripod in the third room to the right: go, therefore, and put it on, and bring the harp, and play to me, my dear."

On which she rank quick with a little cry of delight; and coming again, sat garlanded, incarnadine within the flushing depths of the gold, nor did I send her home to her lonesome yali till the moon, subdued and pallid now from all-night beatitudes, sank down soft within purples, quits of curdling pearl, to the Hesperian realms of her rest.

So sometimes we speak together, she and I, she and I.

* * *
* * *

That ever I should write such a thing! I am driven out from Imbros!

I was strolling in a wood yesterday up to the west—a clear evening, the sun just set, the book in which I have written in my hand, for I had thought of making a sketch of an old windmill to the north-west, to show her. Twenty minutes previously she had been with me, for I had chanced to meet her, and she had come, but had kept darting on ahead after nuts, gathering armfuls of amaranth, nenuphar, red asphodel, till, weary of my life, I had called to her: "Go away! out of my sight," whereupon she, pushing her underlip toward crying, had walked off.

Earthquake destroys the palace -

Well, I was going on in my stroll, when I seemed to feel some quaking of the ground, and before one could count twenty, it was as if the land was bent upon wracking itself to fragments; so in a great scare I set to running, calling in the direction in which she had gone, staggering as on the deck of some laboring craft, tumbling, gathering myself up, running again, the air full of uproar, the land waving like the ocean; and, as I went plunging, little knowing whither, I saw to my left some four roods of forest droop and sink into a ravine which opened to receive them; upon which up I cast my arms, crying out "God! save the girl!", and a minute later rushed out, to my surprise, into open space on a hill-side, whence I could see the palace below, and, beyond it, a wisp of white sea that had the appalling aspect of being higher than the land. Down the hill-side I stumbled, driven by the impulse to flee somewhither, but about half way down was afresh startled by a shrill pattering like musical hail, and in two moments more the palace plunged down with the jangling and clatter of a thousand bells of gold into the bosom of the lake.

Some seconds after this the commotion, having lasted fully ten minutes, commenced to lull. . . . I found her an hour later standing among the ruins of her yali.

* * *

What a thing! Probably every building on the island has been destroyed; the palace-platform, all cracked lies tilted, half-sunken awry into the lake, like an ark stranded, while of the palace itself no trace remains, except a mound of gold-stones emerging above the lake's surface to the south, gone, gone—sixteen years of vanity and vexation. But, from a practical point of view, the direst calamity of all is that the *Speranza* now lies high-and-dry in the village, she having been bodily picked up from the quay by the tidal-wave, and driven bowforemost into a street not half her width; and there now lies, looking huge enough in the little village, wedged for ever, smashed-in at the nip like a match-box, a most astonishing spectacle: her bows forty feet up the street, ten feet above the ground at the stem, rudder resting on the quay, foremast tilted forward, and that bottom which has roamed through seas so remote ambushed in a polychrome of sea-weeds, the old *Speranza;* but, as her steps were there, and by a leap I could catch them underneath and go up hand-over-hand, till I got foothold, this I did at ten the same evening when the sea-water had drained back from the land, leaving everything swampy; she there with me, and presently following me upon the ship. Most things I found cracked into fragments, twisted, disfigured out of recognition; the house-walls themselves displaced a little at the nip; the bow of the cedar skiff smashed in to her middle against the galley; and, but for the fact that the air-pinnace had not broken from her heavy ropings, and one of the compasses still whole, I do not know what I should have done: for those four old boats that had been in the cove have completely disappeared.

I made her sleep on the cabin-floor amid the *débris* of everything, I sleeping high up in a wood to the west, and I write now lying in the long-grass the morning after, the sun rising, though I cannot see him. My plan for today is to cut four logs with the saw, lay them on the ground by the ship, lower the pinnace upon them, roll her down into the water, and by nightfall bid a long farewell to Imbros, which drives me out in this way. Still, I look forward with pleasure to our hour's run to the Mainland, when I shall teach her to steer by the compass, and manipulate liquid-air, as I have taught her to dress, to talk, to cook, to experiment, to write, to think, to live: for she is my creation, this creature, as it were a "rib from my side."

But the "design" of this expulsion, if there are "designs"? and what was it that she called it last night?—"this new going out flom Halan!", this "Haran," it appears, being the place from which "Ablaham" went out, when "called" by God.

* * *

Apparently we felt only the tail of the earthquake at Imbros, for it has broken up Turkey! and we two poor helpless creatures put down here in the theatre of these distractions, it is too bad, for the rages of Nature at present are just amazing, and what it may come to I do not know. When we came to the Macedonian coast in moonlight we sailed along it, and up the Dardanelles, looking out for village, yali, or any habitation where we might put up: but everything wrecked, Kilid-Bahr, Chanak-Kaleh, Gallipoli, Lapsaki in ruins. At Lapsaki I landed, leaving her in the boat, and picked my way a little inward, but soon went back with the news that not even a bazaar-arch was left standing whole, in most parts even the line of the streets being obliterated, for the place had tumbled like a house of dice, and had then been shaken up and jumbled. Finally we slept in a forest on the other side of the strait, beyond Gallipoli, taking our few provisions, and having to wade at some points through morass two feet deep before we arrived at dry woodland.

In this forest the following morning I sat alone—for we had slept separated by half a mile—thinking out the question of whither I should go: my choice would have been to remain either in the region where I was, or to go Eastward; but the region where I was presented no dwelling that I could see; to go any distance Eastward I needed a ship, and of ships I had seen during the night only wrecks, nor did I know where to find one anywhere in this country: I was thus, like her "Ablaham," directed Westward.

In order, then, to go Westward, I first went further Eastward, once more entered the Golden Horn, once more went up those scorched Seraglio steps. Here what the wantonness of man had spared the wantonness of Nature had destroyed, for the few houses that I had left standing round the upper part of Pera I now saw as low as the rest; also the house near the Suleimanieh, where we had lived our first days, to which I now returned as to a home, I found without a pillar standing; and that night she slept under the half-roof of a little funeral-kiosk in the scorched cypress-wood of Eyoub, and I a mile off, at the verge of the forest in which first I saw her.

The following morning, on meeting, as agreed, at the spot of the Prophet's mosque, we passed together through the valley and cemetery of Kassim, by the quagmires up to Pera, all the landscape having to me a twisted unfamiliar aspect. We had determined to employ the morning in searching for supplies among the earthquake-ruins of Pera; and, as I had decided to collect enough in one day to save us further pains for some time, we passed hours in this task, I confining myself to the white house in the park overlooking Kassim, where I had once slept, losing myself amid the obliquities of its floors, roofs, wall-fragments, she going to the Mussulman quarter of Djianghir near, on the heights of Taxim, where were many shops, and thence round the brow of the hill to the French Embassy-house overlooking Foundoucli and the sea, both of us having carpet-bags, and all within the air of that wilderness of break-up that morning a strong, permanent perfume of maple-blossom.

We met toward evening, she quivering under such a load, that I would not let her carry it, but abandoned my day's labor, which was lighter, and took hers, which was quite enough; and we went back westward, prying the while for some shelter from the drenching night-dews of this place, but nothing could find, till we came again, quite late, to her broken funeral-kiosk at the entrance to the immense cemetery-avenue of Eyoub. There without a word I turned from her, leaving her among the wracked catafalques, for I was weary, but, having gone some distance, turned back, thinking that I might take some more raisins from the bag; and, after getting them, I said to her, shaking her little hand where she sat under the roof-shadow on a stone: "Goodnight, Clodagh."

She did not reply promptly: and her reply, to my surprise, was a protest against her name, for a rather sulky, yet gentle, voice came from the darkness, saying: "Am *I* a poisoner?"

"Well," I said, "all right, tell me whatever you like that I should call you, and henceforth I will call you that."

"Call me Eve," says she.

"Well, no," I said, "not Eve, anything but that: for *my* name is Adam, and we do not wish to be ridiculous in each other's eyes; but I will call you anything else that you like."

"Call me Leda," says she.

"And why Leda?" said I.

"Because Leda sounds something like Clodagh," says she, "and you are al-leady in the habit of calling me Clodagh; and I saw 'Leda' in a book, and liked it: but Clodagh is hollible!"

"Well, then," I said, "Leda it shall be, for I like it, too, and you ought to have a name beginning with an 'L'. Good-night, my dear, sleep well, and dream, dream."

"And to you, too, may God give dleams of peace and pleasantness," says she; and I went.

And it was only when I had lain myself on brake for my bed, my head on my caftan, a brook's babbling for my lullaby, and two stars, which alone of the skyful I could spy, for my night-lights, and only when my eyes were already closed toward slumber, that a sudden strong thought wrought and woke me: for I remembered that Leda was the name of a Greek girl who had conceived twins. In fact, I should not be surprised if this "Leda" is the same as "Eve," for all languages were connected at bottom, I have heard of *v*'s interchanging in this way with *b*'s, even with *d*'s, and if *Di*, meaning God, of Light, and *Bi*, meaning Life, and Io*v*e and Iho*v*an and Go*d*, meaning much the same, are all one, that would be nothing astonishing to me, as wi*d*ow and veu*v*e are one; and where it says "truly the Light is Good (*t*ō*b*, *b*on)," this is as if it said "truly the Di is Di." Such, at any rate, is the fatality that tracks me, even in little things: for this Western Eve, or Greek Leda, had twins. . . .

* * *

Well, the next morning we moved through the ruins of Greek Phanar and across the triple stamboul-wall, which still shewed its ivied portal, to make our way, not without climbing, along the Golden Horn to the foot of the Old Seraglio, where I soon came across traces of the railway: and that minute commenced our journey across Turkey, Bulgaria, Servia, Bosnia, Croatia, to Trieste, occupying no day or two as in old times, but four months, a prolonged nightmare, though a nightmare of pleasance, if one may say so, leaving on the memory an immense impression of ravines, ever-succeeding profundities and greatnesses, jungles strange as some moon-struck poet's fantasy, everlasting glooms, and a grieving of unseen rivers, cataracts, and slow cumbered brooks whose bulrushes never behold any ray of sun or moon, with largesse everywhere, secrecies, profusions, the unspeakable, the unimaginable, a savagery most lush and fierce and showy, and valleys of Arcadie, remote mountain-peaks towering, and tarns gnome-guarded like old-buried treasure, and glaciers, and we two human folk pretty small and drowned and lost in all that houselessness, yet moving always through it.

We followed the rails that first day till we came to a train, of which I found the engine good enough, and everything necessary to move it at my hand, but the metals in such a condition of twisted, broken, vaulted, buried mêlée from the earthquake, that, having run some hundreds of yards to examine them, I determined that nothing could be done in that way—a thing that at first threw me into a state like despair for what we were to do I did not know; but after preservering on foot during three days over the track, which is of that large-gauge type of Eastern Europe, I began to see that, deep rusted as it was, there were considerable bits still good, and took heart.

I had with me land-charts and compass, but nothing for taking altitude-observations, for the *Speranza* instruments, except one compass, had all been broken-up by her shock; however, on getting to the town of Silivri, about forty miles from our start, I saw among the ruins of a bazaar-shop a number of brass objects, and found sextants, quadrants, theodolites; two mornings after which we came upon an engine in mid-country, with coals in it, a stream near, the machinery serviceable, as I found after an hour's inspection, having examined the boiler with a candle through the manhole, but red with rust, and the connecting-rod in particular so frail-looking, that, though I had a goat-skin of almond-oil, I felt very dubious: I ventured, however; and, except for some leakage at the tubulure which led the steam to the valve-chest, all went so well, that, at a pressure never exceeding three atmospheres, we travelled nearly a hundred miles before being stopped by a head-to-head block on the line, when we had to abandon our engine. We then continued another nine miles a-foot, I all the time mourning my motor, which I had had to leave at Imbros, and hoping at every townlet to see a whole one, but in vain.

* * *

It was wonderful to see the villages and towns reverting to the earth, already invaded by vegetation, scarcely any more breaking the continuity of "nature," the town now as much the country as the country, and that which is not-man becoming all in all with a certain *furore* of robustness. A whole day among the southern gorges of the Balkan Mountains the train-engine went tearing its way through many a mile of bindweed tendrils, an interminable curtain, burning with flowers of great size, but sombre as the shades of night, rather resembling jungles of Java and the Filipinas; and she that day, lying in the one carriage behind, where I had made her a little yatag-bed from Tatar Bazardjik, continually played the kittur, barely touching the strings, and crooning low, low, in her contralto, everlastingly the same air, over and over anew, crooning, crooning, some moody tune composed out of her own soul's music, just audible to me through the monotony of the engine's slow toiling, until I was drunken with so sweet a woe, my God, a woe that was sweet as swooning, and a dolor that lulled like sleep, and a grief that soothed like peace, so sweet, so sweet, that all that tangle of wood and gloom lost locality and actualness for me, and became nothing but a spell-bound and pensive Heaven for her to moan and lullaby in; and from between my fingers streamed plenteous tears that day, and all that I could keep on mourning was "O Leda, O Leda, O Leda," till my heart was near to break.

The strap of the eccentric of this engine, which was very poor and flaky, suddenly snapped at a pin near five in the afternoon, so I had to stop in a fright; and now that invisible mechanism which had crooned and crooned about my ears in the air, and had followed me whithersoever I went, stopped, too, as down she jumped, calling out: "Well, I had a plesentiment that something would happen, and I am glad, for I was tired!"

Seeing that nothing could be done with the eccentric-strap, I got down, took the bag, and, parting before us the continuous screen, we went pioneering to the left between a rock-cleft, stepping over rocks that seemed negroid with moss-growths: no sight of sky through hundreds of feet of leafage overhead; and everywhere profusions of ferneries burdened with dews, rebellions of dishevelled maidenhairs among mimosas which had a large leaf, with wild vine, white briony, and an odor of cedar, and a soft gushing of waters which informed all that gloaming. The way led upward three hundred feet; and presently, after some windings, and the climbing of five great steps almost regular, yet natural, the gorge opened in a roundish gap, forty feet across, with overhanging crags nine hundred feet on high; and there, behind a screen which fell from the heights, its tendrils defined and straight like a bead-hanging, we spread the store of foods, I opening the fruits, vegetables, meats, wines, she arranging them among the gold-plate, lighting both the spirit-lamp and the lantern, for here it was quite dark. The light revealed behind the screen of tendrils a green cave in the crag, and at the cave's opening a pool two yards wide, black but pellucid, which leisurely wheeled, discharging a streamlet that came out from the cave: and in it I saw four owl-eyed fish, a finger long, loiter, and instigate themselves, and gaze. So there we ate and lingered, until Leda, after smoking a cigarette, said that she would go and "lun," and went, and left me glooming: for she is the sun and the moon and the host of Heaven; I occupying myself that night in making the calendar at the end of this book—for my almanac was lost with the palace—making a calendar, counting the days in my head, but counting them across my thoughts of her.

Then she came again to tell me goodnight, and went down to the train to bed; while I, putting out the light, stooped within the cave, and, spreading my bed beside the rivulet, slept.

He confesses his love – but to sleeping Leda

But an uneasy sleep: for soon I woke; and a long while I lay awake, conscious of a dripping at one spot in the cave, which at a minute's interval darkly splashed, regularly, seeming to grow ever louder, sadder, and the splash was "Leesha," but it became "Leda" to my ears, and it sobbed her name, until I pitied myself, so sad I was. And when I could no longer bear the anguish of the splash and the spasm of its sobbing, I got up to go, soft, soft, lest she should hear in that muteness of the hushed gloom, going more slow, more soft, as I moved more near, a sob stuck in my gullet, my feet leading me to her; till I touched the coach, against which through a long hour I leant my brow, the sob aching within my throat, she all mixed up in my head with the suspended night, and with the elfin hosts in the air that made the silence so vocal to the vacant eardrum, and with that dripping that grieved within the cave; and gradually I turned the handle, heard her breathe in sleep, her head near me, touched her hair with my lips, and near to her ear I said, for she breathed as if in sleep, "Leda, I have come to you, for I could not help it, Leda: and oh, my heart is full of the love for you, for you are mine, and I am yours; and to live with you, till we die, and after we are dead to be near you still, Leda, with my broken heart near your heart, little Leda——"

I must have sobbed, I think: for, as I spoke close at her ear with dying eyes of love, I was startled by a break in her breathing, and in wary haste I closed the door, and quite back to the cave I stole in haste.

And the next morning when we met I thought—but am not now sure—that she smiled singularly: I thought so. She may, she *may*, have heard—— But I cannot tell.

* * *

Twice I was obliged to abandon engines in consequence of forest-tree blockages right across the line, which, do what I might, I could not move, these being the two bitterest incidents of the pilgrimage; and at least twenty times I changed from engine to engine, when other trains obstructed. As for the extent of the earthquake, it is pretty certain that it was universal within the Peninsula, and at many points exhibited superlative violence: for up to the time that we entered upon Servian territory we occasionally came upon stretches of the rails so dislocated, that it was impossible to continue upon them; nor during the whole course did I encounter a house or castle intact; and thrice, where the ground was of a kind to allow of it, I left the rags of metals and made the engine travel the ground till I came upon other metals, when I always contrived to drive it upon them. It was all very leisurely: for not everywhere, nor every day, could I get a nautical observation; and, having at all times to drive at low pressures for fear of tube and boiler weakness, crawling through tunnels, and stopping when darkness came on, we did not advance fast, nor particularly cared to. Once, moreover, for two days, and once for four, we were overtaken by storms of an inclemency so vast, that no thought of travelling entered our heads, our only care being to conceal our cowering bodies as deeply down as possible. Once I passed through a town (Adrianople) doubly devoured, once by the arson of my own arm, and once by Nature: and I made haste to put that place behind me.

Finally, three months and twenty days from the date of the earthquake, having traversed only 900 miles, I let go in the Venice lagoon on the morning of the 10th of September the lateen sail and stone anchor of a Maltese *speronare* which I had found, and partially cleaned, at Trieste; and thence passed up the Canalazzo in a gondola; for I said to Leda "In Venice will I pitch my patriarch tent."

But to will and to do are not the same thing, and still more Westward was I driven: for some of the stagnant canals of this place are now miasmas of pestilence, and within two days I was rolling with fever in the Old Procurazie Palace, she standing in pallid astonishment near me, sickness a novel thing to her; and, indeed, this was my first illness since my twentieth year when I had overworked, and went on a tour to Constantinople. I could not move from bed for a fortnight, but fortunately did not lose my senses, she bringing me the whole pharmacopœia from the shops, from which to choose my medicines; and, divining the cause of this illness, though not a sign of it came near *her*, as soon as my knees could bear me I anew set out, ever Westward, enjoying now a certain luxury in travelling in comparison with that Turkish difficulty, for here were no twisted metals, more and better engines, in the cities as many motors as I chose, and Nature markedly less savage.

I do not know why I did not stop at Verona or Brescia, or some other neighborhood of the Italian lakes, since I was fond of water; but I had, I think, the thought in my head to travel back to Vauclaire in France, where I had lived, and there live: for I thought that she might like those old monks. At all events, we did not remain long in any place till we came to Turin, where we spent nine days, she in the house facing mine; and after that, at her own suggestions, we went on still, passing by train into the valley of the Isère, and then into that of the Rhone, until we came to the old town of Geneva among mountains peaked with snow, the town seated at the head of a lake made in the shape of the crescent moon, and, like the moon, a thing of much beauty and many moods, suggesting a being under the spell of charms and magics. However, with this idea of Vauclaire still in my head, we left Geneva in a motor at four in the afternoon of the 17th of May, I intending to get to the town called Bourg about eight; but by some chance for which I cannot to this hour account (unless the rain was the reason), I missed the road marked in the chart, which should have been fairly level, and found myself on mountain-tracks, unaware of my whereabouts, while darkness fell, and a downpour of rain that had something of a sullen venom drowned us. I stopped often, peering about for château, châlet, or village, but none did I see, though I thrice came upon railway-lines; and not until midnight did I run down a rather steep pass upon the shore of a lake, which, from its apparent vastness in the moonless obscurity, I could only presume to be the great lake once again, three hundred yards to our left being visible through the rain a building apparently risen out of the lake, looking ghostly livid, for it was of white stone, not high, but big, an old thing of complicated turrets (their whiteness roofed with maroon candle-extinguishers), oddities of Gothic nooks, and window-slits, like a fanciful picture. Round to this we drove, drowned as rats, she sighing and bedraggled, to find a spit of land projecting into the lake, on which we left the car, walked forward along it with the bag, crossed a tiny drawbridge, and so got to the islet of rocks on which the castle stands. On finding an open portal, we then went investigating the place, quiet gay at the shelter, everywhere lighting candles which we saw in iron sconces: so that, as the castle is far seen from the shores of the lake, it would have looked to one watching thence a place suddenly possessed and haunted. Having found beds and slept there, the next day we found it to be the Castle of Chillon; and there we remained five happy months, till again, again, Fate overtook us.

* * *

The morning after our coming we had breakfast—our last meal together—on the first floor in a pentagonal room entered from a lower level by three steps, an oak table in it pierced by a multitude of tunnels, worm-eaten, with three chairs having backs two yards high, an oak desk covered still with papers, arras on the walls, three dark oil-paintings, and a grandfather's clock. This room is at the middle of the château, and contains two oriels looking upon the lake, upon an islet containing four trees in a jungle of a river which proved to be the Rhone, upon a snowy town on the slopes which proved to be Villeneuve, and upon the mountains back of Bouveret and St. Gingolph—all having the astonished air of a resurrection just accomplished, everything fresh washed in dyes of azure, ultramarine, indigo, snow, emerald, that fresh morning, so that one had to call it the best and holiest place in the world. These five room-walls, and oak floor, and two oriels, became specially mine, though it was really common-ground to us both, and there I would do many little things, the papers on the desk telling that it had been the *bureau* of one R. E. Gaud, *"Grand Bailli,"* whose residence the place may have been.

She asked me while eating that morning to stay here, and

I said that I would see, though with misgiving: so together we went about the house, and, finding it unexpectedly spacious, I consented to stay, at both ends being suites, little rooms, infinitely quaint and cosy, with heavy furniture Henri Quatre, and bed-draperies; and there are separate, and as it were secret, stairs for exit to each suite, spirals, so we decided that she should have the suite overlooking the length of the lake, the mouths of the Rhone, Bouveret and Villeneuve, and I should have that overlooking the spit of land behind, the drawbridge, the shore-cliffs, and the elm-wood which comes down to the shore, giving a glimpse of the Village of Chillon; and, that decided, I took her hand in mine, and I said: "Well, then, here we stay, under the same roof—for the first time. Leda, I will not explain why, but it is dangerous; so much, that it *may* mean the death of one or other of us: deadly dangerous, my poor girl, believe me, for I know it. Well, this being so, you must never come near my part of the house, nor I near yours. Lately we have been much together, but, then, we have been active, full of purpose and occupation; here we shall be nothing of the kind, I can see: so we must live perfectly separate lives. You do not understand—but things are so. You are nothing to me, really, nor I to you, only we live on the same earth, which is nothing—a chance: so your own food, clothes, everything, you will procure for yourself—perfectly easy—the shores crowded with mansions, castles, towns; and I the same. The motor down there I set apart for your use: I will get another; and I will look you up a boat and fishing-tackle, and cut a cross on the bow of yours, so that you may never use mine. All this is very necessary: you cannot dream, but I know, how much. Do not run any risks in climbing, now, or with the motor, or in the boat . . . Leda . . ."

I saw her under-lip push, and went away in haste, for I did not care whether she cried or not. In that Balkan voyage, and in my illness at Venice, she had become too near and dear to me, my tender love, my dear darling soul; and I said in my heart: "I will be a decent being; I will turn out trumps."

* * *

Under this castle is a sort of dungeon in which are seven pillars, and an eighth half-built into the wall, one of them worn away by some prisoner, or prisoners, once chained to a ring in it, and in this pillar the name "Byron" inscribed—which made me remember that a poet so named had written something about this place; and two days afterwards I actually came upon the poet in a room containing books, many of them English, near the Grand Bailli's *bureau*: so I read the poem, which is named "The Prisoner of Chillon," and found it affecting, the description good—only I saw no seven rings, and where he speaks of the "pale and livid light," he should speak rather of the dun and brownish gloom, for the word "light" disconcerts the fancy here, and of either pallor or blue there is there no sign. However, I was so struck by the horror of man's atrocity to man, as depicted in this poem, that I resolved that she should see it: so went straight to her rooms with the book, and, she being away, ferreted among her things to see what she was doing, finding everything very tidy, except in one room where were a number of prints called *La Mode,* and *débris* of snipped cloth, and medley. When two hours later she came in and I suddenly presented myself, "*Oh!*" she let slip, then fell to cooing her laugh; and now I took her down through a large room stacked with every kind of rifle, with revolvers, cartridges, swords, bayonets—some cantonal magazine—then in the dungeon showed her the worn stone, the ring, the slits in the thickness of the wall, and told all the story of ferocity; while the plashing of the lake upon the rock outside came in with a strange and tragic sound, and her mobile face became all one sorrow.

Then, "How lude and clude they must have been!" cries she with a tremulous lip, her face reddened with indignation.

"Brutes," I said: "it is not surprising if brutes were cruel."

And in the time while I said this she was looking up at me with a smile. "Some others came and set the plisoner flee!"

"Yes," I said, "they did, but—Yes, that was all right, so far as it went."

"And it was a time when men had al-leady become cluel thlough lack of land," says she: "if those who set him flee were so kind when the lest were cluel, what would they have been at a time when the lest were kindly? They would have been just like angels . . . !"

* * *

At this place fishing and rambling were the order of the day, both for her and for me, especially fishing, though a week rarely passed which did not see me at Bouveret, St. Gingolph, Yvoire, Messery, Nyon, Ouchy, Vevey, Montreux, Geneva, or one of the two dozen villages, townlets, towns, that crowd the shores, all very pretty places, each with its charm; mostly I went on foot, though the railway runs right round the forty miles of the lake's length; and one noon-day I was walking through the main-street of Vevey, going on to the Cully-road, when I had an awful shock: for from a shop just in front of me to the right there came a sound—an unmistakable indication of life—a clattering, as of metals rattled together ;whereat my heart bounded into my mouth, I was conscious of becoming bloodlessly pale, and on tip-toe of exquisite caution I stole up to the open door—peeped in—and it was *she,* standing on the counter of a jeweller's shop, her back toward me, her head bent down over a tray of jewels in her hands, which she was rummaging for something. I went "*Hoh!*", for I could not help it: and that whole day, till sunset, we were very dear friends, for I could not part from her, we walking together by voralpen, wood, and shore all the way to Ouchy, she like a creature crazy that day with the bliss of living, rolling in grasses and down flowery slopes, stamping her foot challengingly at me, superb ruler of Earth that she is, then rushing like mad for me to catch her, with laughter, *abandon*, brazen railleries, gaieties of the wild ass's foal on the hills, entangling her loosened hair with the tendrils and blooms of Bacchants, and quaffing, in the passage through Cully, more wine, I fancied, than was right: and the lightning-shocks that shot through me that day, and the rubious revelations of Beauty which my mind's eye sighted, and the pangs of white-hot honey that spanked me, and were too much for me, and made me sick—oh, Heaven, what pen could express any of the recondite realm of things? till, at Ouchy, with a wave of my arm I motioned her back from me, for I was dumb, and weak, and I went away, leaving her there; and all that long night her might was upon me, for she is stronger than gravitation, which may be evaded, than all the forces of nature in combination, and the sun and the moon are nothing compared with her; and when she was no more with me I was like a fish in the air, or like a beast in the deep, for she is my element to breathe in, and I drown without her: so that for hours I lay on that wood-lane mounting to the burial-ground outside Ouchy that night, like a man sore wounded, biting the grass.

What made things horrider for me was her adoption of European clothes since coming to this place. I think that, in her adroit way, she herself made her dresses: for one day I noticed in her rooms some "fashion-plates," with a confusion like dress-making; or she may have been only modifying costumes from the shops, for her Western dressing is not quite like what I remember of the modern style, but is really, I believe, her own goût, nearer resembling the Greek, or the "Empire." At any rate, the airs and graces are not less natural to her than plumage to parrots; and she has changes like the moon, never twice the same, and ever transcending her last phase and revelation: for I could never have imagined anyone in whom *taste* was a faculty so separate as in her, so positive and prominent, like smelling or sight—more like *smelling*: for it is the faculty, half Reason, half Imagination, by which she fore-scents precisely what I will wed exquisitely with what, so that every time I see her I receive the impression of a perfectly novel, completely bewitching, work of Art, the quality of works of Art being to produce the momentary conviction that anything else whatever could not possibly be so good.

Occasionally from my window I would see her in the wood beyond the drawbridge, cool and white in the shade, with her Bible probably or chemistry-book, trailing her train like a court-lady, looking taller than before; and I believe that this new dressing produced a separation between us more complete than it might have been: for especially after that day between Vevey and Ouchy I was careful not to meet her; and the more I noted that she bejewelled herself, powdered herself, embalmed herself in pouncings of nard and sachets of scents, chapleted her head, the more I shunned her. Myself, somehow, had now resumed European dress, and, ah me, was greatly changed, God knows, from the portly monarch-being that had strutted and moaned four years before in the palace at Imbros, so that my manner of being and thought might once more now have been called "modern."

All the more was my sense of responsibility awful: and from day to day it seemed to intensify, a voice never ceasing to remonstrate within me nor leaving me peace, the malediction of unborn billions appearing to menace me; and to strengthen my fixity I would often overwhelm myself, and her, with names of scorn, calling myself "convict," her "lady-bird," asking what manner of man was I that I should dare so great a thing, and, as for her, what was she, to be the Mother of a host?—a butterfly with a woman's brow! and frequently now in my fiercer hours I was meditating either my death—or hers.

Ah, but the butterfly did not let me forget her brow! To the south-west of Villeneuve, between the forest and the river, is a field of gentian, and, returning from round St. Gingolph to the Château one day in the third month, I saw, as I turned a corner in the descent of the mountain, some object floating in the air above the field. Never was I more startled, or more perplexed: for I could see nothing to account for the object soaring there like a great butterfly, though I was soon able to come to the conclusion that she was reinvented *the kite*, and presently sighted her holding the string in the mid-field. Her invention resembles the kind called "swallow-tail" of old.

* * *

But mostly it was on the lake that I saw her, for there we mainly lived, and occasionally there were guilty approaches and *rencontres*, she in her boat, I in mine, both slight clinker-built pleasure-boats of Montreux which I had spent some days in overhauling and varnishing: mine having jib, fore-and-aft mainsail, nad spanker; hers rather smaller, one-masted, with an easy-running lug-sail. It was no uncommon thing for me to sail quite to Geneva, and come back from a seven-days' cruise with my soul inflated and consoled with the lake and its many moods of smiling and darksome, flighty and pensive, dolorous, despairing, tragic, at morning, at noon, at sunset, at midnight: a panorama that never for a moment rested from unrolling its transformations, I sometimes climbing the mountains as high as the goatherd region of hoch-alpen, once sleeping there; and once I was made ill by a two-weeks' horror which I had: for she disappeared in her skiff, I being at the Château, and did not come back; while she was away there was a gale that changed the lake into an angered ocean, and, ah, my good God, she did not come; till at last, half-crazy at the vacant days of care which rolled by and by, and she did not come, I set out upon a wild-goose quest of her—of all hopeless things the most hopeless, for the globe is great—and I did not find her: so after three days I turned back, recognizing that I was mad to search the infinite; and, coming nigh the Château, I saw her wave her handkerchief from the island-edge, for she had divined that I had gone to ferret for her, and was watching for me: and when I took her hand, what did she say to me, the Biblical simpleton? "Oh, you of little faith!" says she; and, since she had adventures to lisp, with the *r*'s liquefied into *l*'s, I was with her that day again.

Once a month perhaps she would knock at my outermost door, which I kept locked when at home, to present me a red trout or grayling sumptuously dressed, which I had not the heart to reject; and exquisitely she does them, all hot and spiced, applying to their preparation that taste which she applies to dress; nor did her luck in angling fail to supply her with the finest specimens, though, for that matter, this lake, with its old fish-hatcheries and fish-ladders, is not stingy in that way, swarming now with the choicest lake-trout, river-trout, red trout, and with salmon, of which last I have brought in one with the landing-net of perhaps forty pounds. As the bottom goes off rapidly from the islands to a depth of nine hundred feet, we did not long confine ourselves to bottom-fishing, but advanced to every variety of manœuvre, doing middle-water spinning with three-triangle flights and sliding lip-hook for jack and trout, trailing with the sail for salmon, live-baiting with the float for pike, daping with blue-bottles, casting with artificial flies; and I could not say in which she became the most carelessly adept, for each soon seemed as old and natural to her as a handicraft learned from birth.

* * *

On the 21st of October I attained my forty-sixth birthday in excellent health: a day destined to end for me in bloodshed and tragedy, alas. I forget now what had caused me to mention the date, long beforehand, in, I think, Venice, not dreaming that she would keep any count of it, nor was I even sure that my calendar was not inaccurate by a day; but at ten in the morning of what I called the 21st, descending by my private spiral in flannels with some trout and par bait, and tackle—I met her coming up, my God, though she had no earthly right to be there; and with her cooing murmur, yet pale, pale, and with a most guilty look, she presents me a big bouquet of flowers.

I was at once thrown into a state of agitation. She was dressed in a frippery of *mousseline* all cream-laced, with short sleeves that hung wide, a diamond at her open chest, the ivory-brown of which looked browner for the powdery blueish-white of her face, where, however, the freckles were not quite whited out, on her feet slippers of silk, pink, without any stockings—a pink pale to fainting, her hair nipped by a ring of gold, and she smelled like Heaven, God knows.

I could not speak. It was she who broke a painful silence, saying, very faint and pale: "It is the day!"

"I—perhaps——" I said, some incoherency like that; and I saw the touch of enthusiasm which she had summoned up quenched by my manner, she presently asking: "I have not done long again?"—looking down, breaking another silence.

"No, no, oh, no," I hurriedly said—"not done wrong again. Only—I could not suppose that you would count up the days. You are . . . considerate. Perhaps—but——"

"Tell Leda?"

"Perhaps . . . I was going to say . . . you might come fishing with me. . . ."

"O, luck!" she went softly.

I was pierced by a sense of my cowardice, my incredible weakness; but I could not at all help it.

So I took the flowers, down we went to the south shore to my boat, from the well of which I threw out some of the fish, arranged the tackle, then the stern-cushions for her, got up the sails; and out we went, she steering, I in the bows, with every possible inch of interval between, receiving delicious whiffs from her of ambergris, frangipane, some imbroglio of fragrances, the morning warm, little whiffs of wind on the water, which was mottled, like water ill-mingled with indigo-wash, we making little headway: so it was some time before I moved nearer to her to get the par for fixing on the three-triangle flight, for I was going to trail for salmon or big lake-trout; and all that time nothing at all was said; but then I said: "Who told you that flowers are proper to birthdays? or that birthdays are of any importance?"

To which she answered: "I suppose that nothing can happen so important as birth; and perfumes were considered ploper to birth, because in the legend the wise men blought spices to the young Jesus."

This *naïveté* was the cause of my immediate recovery: for to laugh is to be saved; and I laughed out, saying: "But you read the Bible too much! You should read the modern books."

"Some I cannot lead," says she: "the people seem to have got so collupted; it makes me shudder."

"Well, now, you see, you come round to my point of view," I said.

"Yes, and no," says she: "they had got so *spoiled*, that is all—seem to have become quite dull-witted—the plainest tluths they could not see. I can imagine that those faculties which aided them in their stlain to become lich, and make the lest poor, must have been gleatly sharpened, while the other faculties withered: as I can imagine a person seeing double thlough one eye, and blind on the other side."

"They didn't *want* to see on the other side," I said: "there were some tolerably clear-sighted ones among them, you know; and these agreed in pointing out how, by changing one or two of their old arrangements of Bedlam, they could greatly better themselves: but they listened with listless ears, or sneered. For they had become more or less unconscious of their misery, especially the rich, so miserable were they—like the man in Byron's 'Prisoner of Chillon,' who, when his deliverers came, was indifferent, for he says:

> 'It was at length the same to me
> Fettered or fetterless to be:
> I had learned to love Despair.' "

"Oh, my God," she went, covering her face a moment, "how dleadful! And it seems tlue—they had learned to love despair, to be even ploud of despair. Yet, all the time, I can see, almost all of them were kind, and clever, too, except in the one eye where habit blinded them flom seeing the stars, as *you* only use one hand, by habit. Such a queer, unnatulal feeling it gives me to lead of those people, I can't desclibe it; their motives seem so slavish, tainted, their life so lopsided—tluly, the whole head was sick, the whole heart faint."

"Quite so," I said; "and observe that this was no new thing: in the very beginning of your Bible you read how God saw that every imagination of man's heart is evil. . . ."

"Oh, but none of that is tlue," she interrupted with a pout—"not tlue of the Polynesians, who, enjoying their land in common, lived in sinless gladness at this garden of God, till white slaves, debased by centulies of slavery, went to pleach to their betters, and to steal flom them—not tlue of you and me, whose hearts are not evil."

To this I answered: "Say yours; as to mine you know nothing, Leda."

The semicircles under her eyes had that morning, as often, a certain moist, heavy, pensive and weary something, as of a strumpet fresh from a revel, very sweet and tender; and, looking softly at me with it, she answered: "Yes, I know my own heart, and it is not evil; not even in the least; and I know yours, too."

"Know *mine!*" I cried, with half a laugh.

"Quite well," says she.

At which cool assurance of hers I was so disconcerted, that I answered not a word, but, going to her, handed her the baited flight, swivel-trace, and line, which she paid out; and I had got back again almost into the bows before I spoke again: "Well, this is news to me: you know all about my heart, it seems. Well, come, tell me what is in it!"

Now she was silent, pretending to be busy with the trail, until she said with her face bent, in a voice that I could just hear: "I will tell you: in it is a lebellion which you think good, but is not good. If a stleam will just flow, neither tlying to climb, nor overflowing its banks, but lunning within its channel whither What leads it leads it, it will leach the sea at last, and lose itself in fulness."

"Ah," I said, "but that counsel is not new—what the philosophers used to call 'yielding to Destiny,' 'following Nature;' and Destiny and Nature, I tell you, often led mankind quite wrong——"

"Or *seemed* to," says she— "for a time: as when a stleam wanders north a little, and the sea is south; but it is bound for the sea all the time, and will wind once more. Destiny never could, cannot yet, be judged, for it is not finished, and our lace should follow whither it points, certain that thlough a maze of curves it conducts the world to God, our Home."

"God our home indeed!" I cried, getting very excited: "girl! you talk speciously, but—Whence have you these thoughts in that——? Girl! you talk of 'our race!' But there are only two of us left? Are you talking at me, Leda? Do not *I* follow Destiny?"

"You?" she sighted, her face bent down: "ah, poor me!"

"What should I do, if I followed it?" I queried with a crazy curiosity.

Her face hung lower, paler, in trouble; and she said: "You would come now and sit near me; you would not be there, you would be for ever near me. . . ."

My good God! I felt my face redden. "Oh, I could not *tell* you . . . !" I cried: "you talk the most disastrous . . . ! you lack all responsibility . . . ! Never, never . . . !"

Her face was now covered with her left hand, her right on the tiller, and bitingly she replied with something of venom: "I could *make* you come—*now*, if I chose; but I will not; I will wait upon my God."

"*Make* me!" I cried: "Leda! How?"

"I could cly before you, as I cly often and often . . . in seclet . . . for my children . . . !"

"*You* do? This is news—*children*—!"

"Yes, I cly. Is not the burden of the world heavy upon me, too? and the work I have to do *vely, vely* gleat? And I cly in seclet, thinking of it. . . ."

Now I saw the push and tortion and shiver of her poor little under-lip, meaning tears, whereupon a flame rose in me that was beyond control, and I found myself in the act of rushing through the boat to catch her to me.

Midway, however, I was saved, when a whisper, intense as lightning, arrested me: "Forward is no escape, nor backward, but *sideward* there is a way!" and before I knew what I was doing, I was in the water swimming.

To the smaller of the islands, two hundred yards away, I swam, rested some minutes, and thence to the Castle. I did not look behind.

* * *

Well, from then till five in the afternoon I thought it all out, lying in my damp flannels on the sofa in the recests beside my bed, where it is dark behind the tattered scrap of arras: and what things I suffered that day, and what depths I sounded, and what prayers I prayed, God knows. What complicated the monstrous problem was this thought in my head: that to kill her would be more clement to her than to leave her alone, having killed myself: and, Heaven knows, it was for her alone that I thought, not at all for myself; but to kill her with my own hands—that was too hard to expect of a poor devil like me, a poor common son of Adam, after all, and never any sublime self-immolator, as four or five of them were. And hours I lay there with brows convulsed in an agony, groaning only this: "To kill her!", thinking sometimes that I should be clement to myself, too, and let her live, and not care, since, after my death, I would not see her suffer, for the dead know not anything. Yet that one or other of us must die was perfectly certain, for I knew that I was on the verge of failing in my oath, and that affairs here had reached a crisis: unless we could make up our minds to part . . . ? putting the width of the earth between us? that concept occurred to me, and in the turmoil of my thoughts it seemed a possibility. Finally, about 5 p.m., I resolved upon something: and I leapt up, went down and across the house into the arsenal, chose a small revolver, fitted it with cartridge, took it upstairs, lubricated it with lamp oil, went down and out across the drawbridge, walked two miles beyond the village, shot the revolver at a tree, found its action accurate, and started back. When I came to the Castle, I walked along the island to the outer end, and looked up: there were her pretty Valenciennes, put up by herself, waving inward before the lake-breeze at one oriel; and I knew that she was in the Castle, for I felt it: and ever when she was within I knew, for I felt her with me, and ever when she was away, I knew, I felt, for the air had a dreadful drought, and a fruitlessness, in it. And I looked up for some minutes to see if she would come to the window, then I called, and she appeared. And I said to her: "Come down here."

* * *

Just here is a rock-path down to the water between rocks mixed with tree-shrubs, three yards long: a path, or a lane, say, for at the lower end the rocks and tree-shrubs reach above one's head. There she had made fast my boat to a little linden: and gloomier now than Gethsemane that familiar boat looked

AJ determined they should be parted, one way or the other.

to my gaze, for I knew very well that I would never enter it more, as I walked up and down the path, awaiting her; and from the jacket-pocket in which lay the revolver I drew a box of matches, took two matches, broke off a bit of one; and both I now held between thumb and forefinger, the phosphorus-ends level and visible, the other ends invisible: and I awaited her, pacing fast, and my brow was brutal as Azrael and Rhadamanthus.

She came, pretty pale, poor thing, and flurried, breathing fast. And "Leda," I said, meeting her in the middle of the lane, and going straight to the point, "we are to part, as you guess—for ever, as you guess: for I see very well that you guess. I, too, am sorry, and heavy in my heart . . . to leave you . . . alone. . . . But it must, aye, be done."

Her face suddenly went as sallow as the dead were, when the shroud was already on, and the coffin had become a commonplace by the bed-side; but, in recording that fact, I record also this, that, accompanying this mortal sallowness, which wretchedly shewed up her poor freckles, was a smile, slightly down-drawn: a smile of steady, of disdainful—confidence.

She did not say anything: so I went on. "I have thought long, and have made a plan—which, however, cannot be effective without *your* consent and co-operation; and the plan is this: we go from this place together—this same night—to some unknown spot, some town, say a hundred miles hence—by train; there I get two motors, and I in one, you in the other, we go different ways; after which we shall never be able, however much we may want to, to rediscover each other in this wide world. That is my plan."

She looked me in the face, smiling her smile; and the answer was not long in coming.

"I will go in the tlain with you," says she with decisiveness: "but where you leave me, there I will stay, waiting till I die, or till my God convert you, and send you back to me."

"That means that you refuse my plan."

"Yes," said she, bending her head with great dignity.

Then I: "Well, you speak, not like a girl, Leda, but like a woman now. But still, reflect a minute . . . O, reflect! If you stayed where I left you, I *should* go back to you, sooner or later: so tell me—reflect, then tell me—do you definitely refuse to part from me?"

Her answer was pretty prompt, cool, and firm: "Yes; I lefuse."

I left her then, walked down the path, came back.

"Then," I now said, "here are two matches between my fingers: be good enough to draw one."

Now she was hit to the heart: I saw her eyes widen to the width of horror—she having read of the drawing of lots in the Bible; knew that this meant death for me, or for her.

But she obeyed without a word after one backward start,

and then a hurried hovering indecision of her fingers over the hand I held out. I had decided that if she drew the shorter of the matches, then she should die; if the longer, then I should die.

She drew the shorter. . . .

* * *

This was only what I should have expected: for I knew that God loved her, and hated me.

But instantly upon the shock of the enormity that I should be her executioner, I formed my resolve: to drop shot, too, in the moment after she dropped shot, so disposing my body, that it would fall half upon her, and half by her, so that we might be close always: and that would not be so bad, after all.

When, in a sudden passion of action, I snatched the revolver from my pocket, she did not move, except her withered lips, which, I think, whispered: *"Not yet . . ."*

And I stood with hanging arm, finger on trigger, looking at her, saw her glance down once at the weapon, then she fixed her eyes upward upon my face: and now that same smile, which had disappeared, was on her lips again, meaning confidence, meaning disdain.

Now I waited for her to move her mouth to say something —to end that smiling—that I might shoot her quick and sudden, and she would not, knowing that I could not kill her smiling: and suddenly my pity and love for her changed into a strange resentment and rage against her, for she was making age-long for me what I was doing for her sake; and the thought came into my mind, "You are nothing to me; if you want to die, you do your own killing; and I will do my own killing:" and without uttering anything, I strode away, left her there.

But I think now that this whole drawing of lots was nothing more than a foolery: I think I *never* could have killed her, smiling or no smiling, for to each thing and life is given a particular strength, and a thing cannot be stronger than its strength, strain as it may: it is so strong, and no stronger, and there an end of the matter.

I strode up to the Grand Bailli's *bureau,* a room about twenty feet from the ground, where, though it was now getting dark, I could see, by peering, the face of a grandfather's-clock which I had long since set going—half-past six; and in order to fix some definite moment for the effort of the mortal act, I said: "At Seven." I then locked the door which opens upon three steps near the desk, and the stair-door, then paced the chamber. As there was not a breath of air here, and I was hot, seeming to be stifling, I tore open my shirt at the throat and opened a mullion-space of one oriel; then at twenty-five to seven I lighted two candles on the desk, and sat to write to her, the gun at my right hand; but I had hardly begun, when I seemed to hear a sound at the three-step door four feet to my left, a sound like a scrape of her slipper; on which I stole to the door to crouch and listen, but now could hear nothing further: so returned to the desk, and set to writing, giving some final directions for her life, telling why I died, how I loved her, more than my own soul, wooing her to love me while she lived, and to live on to please me; but, if she *would* die, then to die near me, though how she was to come into the locked room to die near me I did not stop to enquire; anyway, tears were pouring down my cheeks, when, chancing to glance round, I saw her standing in a ghast posture hardly three feet from my back; and the absolute stealth that had brought and put her there, unknown to me, was like a miracle: for the ladder whose top I saw intruding into the opened oriel I knew well, having frequently seen it in a room below, and, as its length was well over twenty feet, its weight could be no feather's: yet I had been aware of not one hint of its impact upon the window. But there, anyway, she was, wan as a spectre.

In the instant that my consciousness realized her my arm instinctively went out to grasp the weapon; but she, darting upon it, got it before me, flew, and, before I had caught up with her, threw it cleanly between two rungs of the ladder out of the window; upon which I dashed to the window to peer down, thought that I saw it down there near a rock, so away to the stair-door I raced, wrung it open, and down the steps, two together, I pelted to get the gun. I remember being touched by some astonishment that she did not follow me, for somehow I forgot all about the ladder standing there for her to go down on. . . .

But I was reminded of it the moment I arrived at the bottom, before ever I had gone out of the house: for the report of the gun rang out—that crack, my God!—and crying out, "Well, God, it is done!" I stumbled on, to tumble upon her in her blood.

* * *

That night! of fingers quivering with haste, of harum-scarum quests and ferretings, of groans, and appeals to God: for there were no instruments, lint, anæsthetics, nor antiseptics that I knew of in the Château; and though I knew of a house in Montreux where I would find them, the distance was infinite, the time an æon in which to leave her bleeding to death; and, to my horror, I remembered that there was barely enough petrol in the motor, and the store usually kept in the house used up. However, I did it, leaving her there on her bed: but *how* I did it, and lived sane afterwards, is another matter.

If I had not been a medical man, she must, I think, have died: the bullet had broken the left fifth rib, then had been deflected, for I found it buried in the upper part of the abdominal wall; and for a frightfully long time she remained comatose. In which state she still was when I took her to a châlet beyond Villeneuve, three miles away on a mountain-side, a homely, but very salubrious place which I knew, imbedded in boscage: for I was desperate at her long collapse, and had hope in that upper air. I did not sleep, only nodded and tottered, and there after two more days she opened her eyes, and smiled with me.

It was then that I said to myself: "This is the noblest, sagest, and also the most lovable, of the beings that God has made: and since she has won my life, I will live. . . . But at least, to save myself, I will put the broadest ocean that there is between her and me, for the honor of my race, being the last, and to turn out trumps. . . ."

Thus, after only fifty-five days at the châlet, were we forced still Westward.

* * *

I wished her to remain at Chillon, intending, myself, to make for the Americas, whence any impulse to slip back to her could not quickly be fulfilled; but she refused, saying that she would come with me to the coast of France: and I could not say her no.

And at the coast after thirteen days we turned up, three days before the New Year, having traversed France by both steam and petrol traction.

To Havre we came—infirm of purpose that I was: for deep in my heart was the secret, hidden away from my own upper self, that, she being at Havre, and I at Portsmouth, we could still speak with each other.

We came humming into that dark town of Havre in a motorcar about ten in the night of the 29th of December, a bitter bleak night, she, it was clear, poor thing, cramped with cold; and, as I had some recollection of the place, for I had been there, I drove to the quays, near which I stopped at the *Maire's* house, a palatial place overlooking the sea, in which she slept, I occupying another near.

The next day I was early astir, searched in the *mairie* for a map of the town, and could thus locate the Telephone Exchange; then to the *Maire's* house, which I had fixed upon to be her home, where I found the telephone in an alcove adjoining a *salon* Louis Quinze; and, fearing any weakness, I connected with the transmitter-circuit some new cells from the accumulator-room at the Exchange; which done, I went down among the ships by the wharves, fixed upon the first old tub that seemed sea-worthy, broke open a shop, procured some buckets of oil, and by three o'clock had tested and prepared my ship—a day of deathbeds drenched in drizzling, chill. I then returned to the *mairie*, where for the first time that day

He parts from Leda at Havre

I met her, and heavy was her soul in her; but when I broke the news that she would be able to talk to me, every day, all day, first she was all surprise and uncertainty, then her eyes turned white to the skies, then she was skipping like a kid; after which we lingered together three hours, going about the town, fetching home stores of whatever she might want, until I saw darkness coming on, and we passed down to the ship.

And when those old screws awoke and moved, bearing me toward the Outer Basin, I marked her standing there darkling on the Quai through heartrending greyness of inclemency, and, ah, God, the gloomy underlook of that gaze, the piteous push of that lip, then the burying of that face! My heart broke, for I had not given her one little kiss of good-bye; and she had been so good, quietly acquiescing, like a good wife, not attempting to force her presence upon me in the ship; and I went and left her there, all widowed, solitary on a continent, blinking after me: and I steered out to the bleak and dreary fields of the sea.

* * *

Arriving at Portsmouth the next morning, I made my residence in the first house in which I found an instrument, a spacious dwelling facing the Harbor Pier, then hurried round to the Exchange, which is on the Hard near the Docks, a red building with facings of Cornish moorstone, a bank on the ground floor, and the Exchange on the first. Here I plugged her number on to mine, ran back, rang—and, to my great thanksgiving, heard her speak. (This instrument, however, did not prove satisfactory even when I had put-in another battery, and at last I put a bed into the middle room at the Exchange, with stores, and here have taken up my residence).

I believe that she lives and sleeps under the instrument, as I here live and sleep, sleep and live, under it; and, my instrument being near one of the beach windows, I, hearing her, can look out toward her over the field of the sea, yet not see her, and she, too, looking out over the sea toward me, can hear my speech coming out of the deeps of nowhere, but see me not.

* * *

I this morning early to her: "Good morning! Are you there?"

"Good morning! No: I am there," says she.

"Well, that was what I asked—'are you there?' "

"But I am not here, I am there," says she: "the paladox of the heart!"

"The what?"

"The paladox!"

"But still I do not understand: how can you be both there and not there?"

"If my ear is here, and elsewhere?" says she.
"An operation?"
"Yes!" says she.
"What doctor?"
"A special one!" says she.
"Ear-specialist?"
"Heart!" says she.
"And you let a heart-specialist operate on your ear? How are you after it?"
"Happy but for a sigh. And you?"
"Quite well. Did you sleep well?"
"Except when you lang me up at midnight. Have had such a dleam . . ."
"What?"
"Dleamed that I saw two little boys of the same age—only I could not see their faces—playing in a wood . . ."
"Ah, I hope that one of them was not named Cain, my poor girl."
"No! neither of them! Suppose I tell a stoly and say that one was named Caius and the other Tibelius, or one Charles and the other Herbert?"
"Ah, well . . . So what will you be doing today?"
"It is a lovely day. . . . Have you nice weather in England?"
"Very."
"Well, at eleven I will go out and gather Spling-flowers in the park, and cover the *salon;* then I will start upon antimony, for I finished arsenic yesterday. . . . Wouldn't you like to be here to do it with me?"
"Not I."
"You would!"
"Why should I? I like England."
"But Flance is nice, too: and Flance wants to be fliends with England, and is waiting, oh, waiting, for England to come over, and be fliends. Couldn't some *lapplochement* be negotiated?"
"Good-bye. This talking spoils my morning smoke. . . ."
So we speak together across the sea, my God.

* * *

On the morning of the 8th of April, when I had been separated thirteen weeks from her, I boarded several ships in the Inner Port, a lunacy in my heart: and I selected what looked like a fast boat, one of the smaller Atlantic "steamers," called the *Stettin,* which seemed to need the least toil in oiling, &c., in order to fit her for the sea: for the boat in which I had come to England was a tub, and I pined for the wings of a dove, that I might fly away to her, and be at rest.

With flustered hands I labored that day, and I should think that I was of the color of ashes to my lips. By half-past two I was finished; and by three was coasting down Southampton

Sailing back to Leda, but he stops himse

Water by Netley Hospital and the Hamblemouth, having said not one syllable at the telephone about going, nor to my own guilty soul a syllable: but in the depths of my being I felt this fact, that this must be a 35-knot ship, and that, if driven hard, she would go 30 against the drag of the garment of seaweed which she trailed; also that, Havre being 120 knots away, at 7 p.m. I should be on its quay.

And when I was away, and out on the bright and breezy sea, I howled to her, crying out *"I am coming!"*, and I knew that she could hear, and that her heart leapt to meet me, for mine leapt, too, and felt her answer.

The sun went low; it set. I was tired of the day's labor, of standing in the breeze at the high-set wheel, could not yet see the coast of France, and a thought smote me: and after a quarter of an hour I threw the boat's bows round, my face screwed with pain, God knows, like prisoners whose fingers were ground betwixt screws, and their body drawn out to tenuous length, and their flesh pinched with pincers; and I fell upon the floor of the bridge contorted with anguish: for I could not go to her. But after a time that paroxysm passed; and I rose up sullen and resentful, to resume my place at the wheel, and steer again for England, a fixed resolve now in my breast; and I said "No, no more: if I could bear it, I would . . . but if it is impossible, how can I? Tomorrow night as the sun sets—without fail—so help me, God—I kill myself."

* * *

So it is finished, my good God.

In the morning of the next day, the 9th, I having come back to Portsmouth about eleven the previous night, when I bid her "Good morning," she said "Good morning," and not another word. I said: "I got my hookah-bowl broken last night, and shall be trying to mend it today."

No answer.

"Are you there?" I said.

"Yes," says she.

"Then, why don't you speak?" I said.

"Where were you yesterday?" says she.

"I went for a cruise in the basin," I said.

Silence for three minutes; then she: "What is the matter?"

"Matter?" I said.

"Tell me!" she says—with such an intensity and rage, as to make me shudder.

"Nothing to tell, Leda!"

"But how can you be so *cluel?*" she cries.

There was anguish in that cry: and the thought took me then, how, on the morrow, she would ring, and have no answer; and she would ring again, and have no answer; and she would ring all day, and ring, and always would ring, with white hair flowing and the eye-balls of frenzy, battering re-

Leda says there's another purple cloud

proaches at the doors of a universe which would howl back everlastingly to her howls only the howl of its soundlessness: and for very pity, my God, I could not help sobbing to myself "May God pity you, woman!"

I do not know if she heard: she must, I think now, have heard; but no reply came; and there I, shivering like the sheeted dead, stood waiting for her next utterance, waiting long, dreading, hoping for, her voice, thinking that, if she sobbed but once, I should drop dead there where I stood, or eat my tongue through, or shriek the laughter of distraction; but when at last, after some forty minutes or more, she spoke, her voice was perfectly firm and calm. She said: "Are you there?"

"Yes," I said, "yes, Leda."

"What was the color," says she, "of the poison-cloud which destloyed the world? Purple, was it not?"

"Yes, purple, Leda," I said.

"And it had a smell like almonds, did it not?" says she.

"Yes," I said, "yes."

"Then," says she, "there is *another* eluption. Evly now and again I seem to scent whiffs like that . . . and there is a vapor in the East which glows—purple it is . . . see if you can see it. . . ."

I flew across the room to an east-window, threw up the sash to look; but, the view being barred by the back of a warehouse, I rushed back, gasped to her to wait, rushed down the two stairs, and out on the Hard ran dodging wildly about, seeking a purview to the East; till finally I ran up the dockyard, behind the storehouses, to the Semaphore, to arrive at the top panting for life; and now I looked abroad, but only to behold all the heavens cloudless, save for a bank of cloud to the northwest, the sun blazing in a space of azure pallor; so back anew I flew, to tell her: "I cannot see it . . . !"

"Then, it has not tlavelled far enough to the north-west yet," she said.

"My wife!" I cried: "you are my wife now!"

"Am I?" says she: "at last? . . . But shall I not die?"

"No! you can escape! My home! My heart! If only for an hour, then death, just think, on the same couch, for ever, heart to heart—how sweet!"

"Yes! sweet! . . . But how escape?"

"It travelled slowly before. . . . Get quick into that boat under the crane—you have seen me turn-on liquid air—that handle under the dial; get oil from that shop next to the clock-tower, and toss it over everything rusted—only spend no time; you can steer by tiller and compass, well, the wheel is the opposite, the course North-East by North—I meet you on the sea—go now——"

I was wild with bliss. I thought that I should take her between my arms, and have the freckles against my face, and taste her short firm-fleshed upper-lip, and moan upon her, and whimper upon her, and mutter upon her, and say "my wife;" and even when I knew that she was gone from the telephone, I still stood there, hoarsely calling after her "My wife! My wife!"

* * *

I flew down, all confident, to where the boat lay moored that had borne me the day before, for, as her joint speed with the speed of Leda's boat would be forty knots, in three hours we must meet; nor had I the least fear of her ceasing to live ere our meeting: for, apart from the gradualness of the vapor's progress that first time, I foretasted and trusted my love, that she would surely come, and not fail, as dying saints foretasted and trusted eternal life.

I was no sooner on boar the *Stettin* than her engines were straining under what was equivalent to "forced draught;" and, although on the day before it would have surprised me at any moment while I drove her to be carried to the clouds in an explosion from her rusted tanks, this day such an apprehension never crossed my mind, for I knew that I was immortal till I saw her.

The sea was quite placid, as on the previous day, and appeared placider, the skies brighter, and there was a flightiness of laughter in the feet of the breezes that frilled the sea in dashing dark patches, like *frissons* of tickling; and I thought that the morning was a genuine marriage-morning, and remembered that it was a Sabbath; and sweet smells our wedding would not lack of almond and peach, though, looking eastward, I could see no blush of any purple cloud, but only whirls of chiffon under the sun; and it would be an eternal wedding, for one day in our sight would be as a thousand years, and our thousand years of delight one day, since in the evening of that eternity death would visit us, sweetly to lay its finger on our sluggard lids, and we'd die tired of delight; and all manner of dancings and singings—fundango and glee of galliard, corantoes and the solemn gavotte—were rampant in my heart that happy day; and, in running by the chart-house to the bridge, I spied under the table a roll of old flags, and presently they were flying in an arc of gala from the main; and the sea rumpled in a tract of tumbling milk behind me; and I hasted homeward to meet my heart.

* * *

No purple cloud could I observe, as on and on, for two hours, I tore southward; but at hot noon, on the port beam, I spied through the glass across the water something else that moved; and it was you who came to me, O, Leda, my spirit's breath!

When I bore down upon her, waving, soon I saw her stand

like the ancient mariner, but in muslins that fluttered, at her wheel on the bridge—one of those little Havre-Antwerp craft, high in the bows—and she waved a little white thing, until I could spy her face, her smile, when I called to her to stop, in a minute stopped myself, and by happy steering came with headway which failed to a slight crash by her side; then ran down the steps to her, led her up; and on the deck, without saying anything, I fell to my knees before her, and I bowed my brow down, down, to the floor, with obeisance, and I adored her there as Heaven.

And we were wedded: for she, too, bowed the knee with me under that jovial sky; and under her eyes were the moist semi-circles of fatigue, dreamy, pensive, so dear and wifish; and God was there, and saw her kneel: for He loves the girl.

Then I got the two vessels apart, and there they rested some yards separated through the day, we two being in a main-deck cabin, where I had locked a door, so that no one might come in to be with my love and me.

* * *

I said to her: "We will fly west to one of the Somersetshire coal-mines, or to one of the Cornwall tin-mines, where we will barricade ourselves against the cloud, and provision ourselves for months, for it is quite practicable, we have plenty of time, and no crowds to break down our barricades—and there in the deep we will live sweetly, till the disaster is overpast."

And she smiled, drew her hand across my face, said: "No, no; don't you tlust in my God? do you think He would leally let me die?"

For she has appropriated the Almighty God to herself, naming Him "*my* God," aye, and she generally knows what she is saying, too: and she would not fly the cloud.

And I am now writing three weeks later at a little place called Château-les-Roses, and no poison-cloud, nor any sign of any poison-cloud, has come: and this I do not understand.

It may be that she conjectured that I was on the point of destroying myself . . . she may be capable . . . But no, I do not understand, and shall never ask her.

But *this* I understand: that it is *the White* who is Master here; that though He wins but by a hair, yet He wins: and since He wins, dance, my heart.

I look for a race that shall resemble its Mother: nimble-witted, light-minded, pious—like her; all-human, ambidextrous, ambicephalous, two-eyed—like her; and if, like her, they talk the English language with all the *r*'s turned into *l*'s, that will be nice, too.

They will be fruit-eaters, I suppose, when the meat now about is eaten up; but it is not known that meat is good for men; and, if it is really good, then they will *invent* a meat: for they will be *her* sons, and she, to the furthest circle within which the organ of woman's wit is ordained to orbit, is, I swear, all-wise.

There was a "preaching" man—a Scotchman he was, named Macintosh—something like that—who said that the last end of Man shall be well, and very well; and she says the same: and the agreement of these two makes a truth. And to that I now say: Amen, Amen.

For I, Adam Jeffson, parent of a race, hereby lay down, ordain, and decree for all time, perceiving it now: That the one motto and watchword proper to the riot and odyssey of Life in general, and in especial to the race of men, ever was, and remains, even this: "Though He slay me, yet will I trust in Him."

THE END